One Earth,
Four or Five Worlds

Some other books by Octavio Paz

The Labyrinth of Solitude
The Other Mexico
Alternating Current
The Bow and the Lyre
Children of the Mire
Conjunctions and Disjunctions
Selected Poems

Octavio Paz

One Earth, Four or Five Worlds

Reflections on Contemporary History

Translated by Helen R. Lane

A Harvest/HBJ Book
HARCOURT BRACE JOVANOVICH, PUBLISHERS
San Diego   New York   London

Library of Congress Cataloging in Publication Data
Paz, Octavio, 1914–
One earth, four or five worlds.
Translation of: Tiempo nublado. With 3 added essays.
1.   History, Modern—1945–   —Addresses, essays, lectures.
I.   Title.
D842.5.P3913   1985   909.82   85-777
ISBN 0-15-169394-3
ISBN 0-15-668746-1 (Harvest/HBJ : pbk.)

Designed by Francesca M. Smith
Printed in the United States of America
First Harvest/HBJ edition 1986
A B C D E F G H I J

Contents

Part Two: As Time Goes By

Preface

It is not without hesitation that I have decided to gather together the series of articles on the recent past which I published in a number of Spanish and Latin American newspapers during the early months of 1980. I have eliminated many pages, some because they were too topical and others because events have overtaken them. I have likewise modified, rectified, and at times expanded certain passages. For this English edition, moreover, I have added three sections that did not appear in *Tiempo Nublado*.

Despite all such changes, I have no illusions about the imperfections and limitations of these writings. I am not a historian. My passion is poetry and my occupation literature; neither gives me the authority to pass judgment on the agitations and upheavals of our era. Naturally, I am not indifferent to what is going on in the world (who can be?), and I have written articles and essays about the contemporary scene, though always from a point of view that I might call eccentric or simply marginal, I am not sure which—but never, in any case, from a perspective firmly rooted in the certainties of an ideology with encyclopedic pretensions such as Marxism, or in the immutable truths of religions such as Christianity and Islam. Nor have I written from the center, real or supposed, of history—

New York, Moscow, or Peking. I do not know whether these commentaries contain valid interpretations or reasonable hypotheses; I do know that they express the reactions and sentiments of an independent writer from Latin America confronting the modern world. They are not theory but testimony.

Like the ancient Mayas, who had two ways of measuring time, the "short count" and the "long count," French historians have introduced the distinction between "long duration" and "short duration" in historical processes. The first designates the broad rhythms that, through modifications that are imperceptible at first, alter old structures and create new ones, thus bringing about gradual but irreversible social changes. Examples include increases and decreases in population, which are still not entirely explained; the evolution of sciences and technologies; the discovery of new natural resources or their gradual exhaustion; the erosion of social institutions; the transformation of mentalities and sensibilities. . . . "Short duration" is the domain par excellence of the event: falling empires, nascent States, revolutions, wars, presidents resigning, assassinated dictators, crucified prophets, sanctimonious crucifiers, etc. History is often compared to a woven fabric, the work of many hands; without deciding on a pattern beforehand, and without knowing exactly what they are doing, these hands weave threads of every color together until a succession of figures, at once familiar and enigmatic, appears on the loom. From the point of view of "short duration," the figures are not repeated: history is constant creation, novelty, the realm of the unique and singular. From the standpoint of "long duration," repetitions, ruptures, and renewals can be perceived: rhythms. Both visions are true.

The majority of the changes that we have experienced belong, of course, to "short duration," but the most significant of them are related, directly or indirectly, to "long duration." In the last ten years historical rhythms that have been at work for more than two centuries have finally become visible. Al-

most all of them are terrifying: population growth in the
underdeveloped countries; the drain on sources of energy; the
contamination of the atmosphere, the seas, the rivers; the
chronic illnesses of the worldwide economy, which passes in
cycles from inflation to depression; the expansion and multi-
plication of ideological orthodoxies, each one with pretensions
to universality; and finally, the bleeding wound of our societies: State terror and its counterpart, the terrorism of bands of
fanatics.

"Long duration" gives us the feeling that we are before a
historical landscape—that is to say, before a history that ex-
hibits the immobility of nature. A misleading impression: nature also moves and changes. The changes of "short duration"
register on this apparently immobile background like phenomena that alter the physiognomy of a place: the coming and
going of light and darkness, noon and twilight, rain and storms,
the wind that blows the clouds before it and raises whirls of
dust.

I have divided this essay into two parts. The first contains
five chapters, on the changes of opinion and outlook in the
nations of the Old World; the crisis of imperial democracy in
the United States; its counterpart, the crisis of the system of
bureaucratic domination in Russia; the revolt of particularisms, above all in the countries on the periphery; and modernization, its dangers and difficulties. In this first part I have
reduced allusions to the situation in Latin America to a minimum—except in the final chapter—because I expand upon this
theme in Part Two.

Part One

Threatening Skies

I

A Bird's-eye View of the Old World

From Criticism to Terrorism

Around 1960 a series of civic upheavals began that made the
West tremble. Contrary to the predictions of Marxism, the crisis was not an economic one, nor was its central protagonist
the proletariat. It was a political crisis, and, even more, a moral
and spiritual one; its agents were not workers but a privileged
group: students. In the United States the rebellion of the young
played a decisive role in discrediting American policy in Indochina; in Western Europe it shattered the credibility and prestige of governments and institutions, if not their power. The
youth movement was not a revolution, in the strict sense of
the word, although it made the language of revolution its own,
nor was it a revolt; it was a rebellion, in the meaning I have
given that term in other writings.[1] It was the rebellion of a
segment of the middle class and a real "cultural revolution,"
in precisely the sense that the so-called one in China was not.
The extraordinary freedom of mores in the West, above all in

---

1. See *Alternating Current* (1973; original Spanish edition, *Corriente alterna*,
1967) and the chapter "Inicuas simetrías" ["Iniquitous Symmetries"], in *Hombres
en su siglo* [Men in their Time], 1984.

the sexual realm, is one of the consequences of the moral insurrection of the young in the sixties. Another is the progressive weakening of the notion of authority, be it governmental
or paternal. Previous generations had seen the cult of the terrible father, adored and feared: Stalin, Hitler, Churchill, De
Gaulle. In the decade of the sixties, an ambiguous image, alternately angry and orgiastic—the Sons—displaced the satur-
nine Father. We went from the glorification of the solitary old
man to the exaltation of the juvenile tribe.

Even though the disturbances in the universities shook the
West, neither the Soviet Union nor the communist parties in
the various countries exploited them or attempted to channel
them. On the contrary: they denounced these disturbances as
anarchic, decadent, petit-bourgeois movements, manipulated
by *agents provocateurs* of the right. The hostility of the Soviet
hierarchy is understandable: the rebellion of the young was a
libertarian movement and a passionate, total critique of the
State and of authority as much as it was an explosion against
the capitalist consumer society.

The following decade saw the emergence, and the recognition in the West, of dissidents in the Soviet Union and in the
other "socialist" countries. This has left its mark on the contemporary intellectual conscience, and its moral and political
consequences will be felt more and more deeply, not only in
Europe but in Latin America as well. For the first time, the
dissidents of the Russian Empire managed to make themselves
heard by European intellectuals; until just a few years ago,
only a few marginal groups—anarchists, Surrealists, former
Marxists and communist militants who had shed their cas-
socks—had dared to describe bureaucratic socialism as what it
really is: a new, more total, and merciless system of exploita-
tion and repression. Today no one dares defend "real socialism" as before, not even the members of that species we call
"liberal intellectuals." Gide's criticisms in 1936 and Camus's
more penetrating ones in 1951, prior to the revelations of the

dissidents, seem timid; Trotsky's analyses insufficient; and even Souvarine's descriptions pale, though the latter, some forty years ago, was the first to understand the true nature of the Russian regime.

During the decade of the seventies, no movement of moral and political self-criticism appeared in the West comparable to that of the dissidents of the "socialist" countries—a notable contrast, yet one to which no one, as far as I know, has given serious thought. This is strange, since from the sixteenth century on, criticism has accompanied Europeans in all their undertakings and adventures, at some times in the form of confession and at others in the form of remorse. The modern history of the West begins with the expansion of Spain and Portugal in Africa, Asia, and America; almost at once, denunciations of the horrors of the Conquest make their appearance, and descriptions—often wonderstruck—of indigenous societies are written. On the one hand, Pizarro; on the other, Las Casas and Sahagún. At times the conquistador himself is, in his own way, an ethnologist (Cortés). The guilt feelings of the West go by the name of anthropology, as Lévi-Strauss has pointed out: a science born at the same time as the European imperialism that it has outlived.

In the first half of the twentieth century the critique of the West was the work of its poets, its novelists, and its philosophers. Their criticism was unusually violent and lucid. The rebellion of the young in the sixties took up the same themes and lived them as passionate protest. The youth movement, admirable on more than one account, oscillated between religion and revolution, eros and utopia.

Then, as suddenly as it had begun, it was over. The rebellion of the young flared up when no one was expecting it and faded away just as unexpectedly, a phenomenon that our sociologists are still unable to explain. A passionate negation of the values reigning in the West, the cultural revolution of the sixties was the offspring of criticism, but, strictly speaking, it was

not a critical movement. By that I mean that the protests, dec-
larations, and manifestoes of the rebels set forth no ideas and
concepts not already found in the philosophers and poets of
the immediately preceding generations. The novelty of the rebellion was not intellectual but moral; the young did not discover other ideas—they lived, with passionate intensity, those
they had inherited. In the seventies, rebellion died away and
criticism fell silent. Feminism was an exception, but this movement began much earlier and will surely continue for several
decades more. It is a process that belongs to the realm of the
"long count." Although it has lost some of its impetus in the
last few years, it is a phenomenon destined to endure and
change history.

The heirs of the youthful rebels have been the terrorist bands.
The West ceased to have critics and dissidents; the minorities
in the opposition turned to clandestine action. This was the
reverse of Bolshevism: incapable of taking over the State and
instituting ideological terror, the activists have been caught up
in the ideology of terror. A portent of the times: as terrorist
groups grow bolder and more intransigent, the governments
of the West become more timid and more hesitant. Is skepticism on the part of governments the only possible answer to
the fanaticism of the terrorists? The frankest justification of the
necessity of the State was Hobbes's: "Since the condition of
man is a condition of war of everyone against everyone," men
have no other recourse save to yield part of their freedom to a
sovereign authority capable of ensuring the peace and tran-
quillity of each and all. Yet even Hobbes conceded that "the
condition of subject is a miserable one." A great contradiction:
to lovers of freedom the erosion of the authority of government in the countries of the West ought to be a cause for re-
joicing; the ideal of democracy may be defined succinctly as a
strong people and a weak government. Yet the situation sad-
dens us because the terrorists seem bent upon proving Hobbes
right. Not only are their methods reprehensible; their ideal is
not freedom but the establishment of sectarian despotism.

However pernicious the actions of these groups, the real evil of liberal capitalist societies lies not in them but in the predominant nihilism. It is a nihilism poles apart from that of Nietzsche: we are not confronted with a critical negation of established values but with their dissolution in a passive indifference. We would be closer to the mark to call it hedonism: the temper of the nihilist is tragic; that of the hedonist, resigned. This hedonism, moreover, is very far removed from that of Epicurus: it does not dare to look death in the face, and it is not a form of wisdom but a surrender. At one extreme it is a kind of gluttony, an insatiable craving for more and more; at the other it is dereliction, abdication, cowardice in the face of suffering and death. Despite the cults of sports and health, the attitude of the masses in the West implies a lowering of vital energy. A person lives for a greater number of years now, but they are empty, hollow years. Our hedonism is a hedonism for robots and wraiths. Viewing the body as merely a mechanism leads to the mechanization of pleasure; in its turn, the cult of the image—movies, television, advertising—gives rise to a sort of generalized voyeurism that converts bodies into shadows. Our materialism is not carnal: it is an abstraction. Our pornography is visual and mental; it exacerbates loneliness and verges at one extreme on masturbation, at the other on sadomasochism—elaborate solitary exercises, at once bloody and spectral.

The spectacle of the West today would have fascinated both Machiavelli and Diogenes, for quite different reasons. Americans, Europeans, Japanese have managed to overcome the postwar crisis and create a society that is the richest and most prosperous in human history. Never have so many had so much. Another great achievement is tolerance—a tolerance not only toward ideas and opinions but also toward mores and inclinations. Yet these material and political gains have not been matched by the attainment of a loftier wisdom or a more profound culture. The spiritual panorama of the West is desolating: cheap tastes, triviality, shallowness, rebirth of superstition, degradation of eroticism, pleasure enrolled in the service of the communications media. But terrorism is not a critique of this situation: it is one of its symptoms. To the somnambulism of society, mechanically spinning round and round the endless production of objects and things, terrorism offers in opposition a frenzy no less somnambulistic but more destructive.

It is by no means an accident that terrorism has been most rampant in Germany, Italy, and Spain—the three countries where the historical process of modern society (the transition from the absolutist State to the democratic one) has been interrupted more than once by despotic regimes. In all three, democracy is a recent institution. The national State—the necessary complement of the evolution of Western societies toward democracy—came into being at a late date in Germany and Italy. The case of Spain was quite the contrary, but the results have been similar: the different peoples living side by side in the Iberian Peninsula were immobilized, from the sixteenth century on, in the straitjacket of a centralist, authoritarian State. This does not mean, naturally, that the Germans, the Italians, and the Spaniards are doomed to terrorism by some kind of historical defect. As democracy and federalism are strengthened (and with them the national State), terrorism will decline. Actually, it has already disappeared in Germany. Nor is one venturing too far to conjecture that it will diminish in Spain as well. It will not be repression by the government but the establishment of local and regional freedom and autonomy that will bring an end to Basque terrorism there. The ETA is doomed to die, not suddenly, but as a consequence of a gradual but inexorable isolation. Since it has ceased to represent popular aspirations, growing isolation will lead it to the worst form of violence: political suicide. The process will be slow but irreversible.

The activities of the Italian terrorists have been, more than anything, the result of the crisis of the State, which in turn has been the consequence of the double paralysis of the two great parties, the Christian Democrats and the Communists. The government goes around in circles, not getting anywhere, because the party in power, the Christian Democrats, no longer has any program beyond maintaining the *status quo*. It does not govern—or, rather, it has reduced the art of governing to a conjurer's trick, in which what counts is dexterity, a mastery of the art of compromise. As for the Communist Party, it does not know which way to turn. It has rejected Leninism but does not dare embrace democratic socialism wholeheartedly. It wavers between Lenin and Kautsky without having found its own proper course yet. Though Italian political life is an extremely agitated affair, nothing happens. Everybody scurries around yet remains in precisely the same place. Nor is the cold, obtuse anger of the terrorists and their pedantic teachers a way out. That is why they have failed. But the underlying problem nonetheless remains: Italy suffers from the absence of a democratic socialism. The Italian communists, the most flexible and intelligent in Europe, have set out to fill this gap, but they have not succeeded because of their historical genealogy. Their long association with Russian bureaucratic despotism is a sort of original sin that as yet no democratic baptism has been able to wash away.

And the case of Northern Ireland? As I see it, we are dealing here with a very different phenomenon. Irish terrorism was born of the explosive conjunction of two elements: a nationalism steeped in religious fervor, and the second-class status forced upon the Catholic minority. The history of the twentieth century has confirmed something well known to all the historians of the past, something our ideologies have stubbornly ignored: the strongest, fiercest, most enduring political passions are nationalism and religion. In the case of the Irish, the union between the two is inextricable. Unlike the Basques, who do not want to unite with anyone except one another, the Catholics of Ulster feel that they are part of the Republic of Ireland. But Irish Catholic nationalism is one thing and the IRA another. Two circumstances conspire against that organization: first, the Catholics are in the minority in Ulster; second, both the IRA's methods and its political program (a "socialism" of the Arab or African variety) have lost it friends and supporters in the Republic of Ireland as well as in Northern Ireland.

The assassination of Lord Mountbatten was condemned by many whose sympathies lay at first with the terrorists. As the IRA becomes radicalized, it isolates itself. But this is a problem that cannot be resolved by armed violence; what is required is a formula that will satisfy, at least in part, the aspirations of the Catholic minority. The situation bears more than one similarity to that of the Palestinians and Israelis—how to satisfy the contradictory and exclusive, though equally legitimate, aspirations of two communities. Unfortunately, there is no Solomon in sight.

The Jacobin Heritage and Democracy

A phenomenon of inverse symmetry: the evolution of the great European communist parties (in Italy, France, and Spain) has taken a diametrically opposite direction from that of the terrorists. As extremist groups step up the violence of their tactics, the communists draw closer to the methods and programs of the traditional democratic parties. If the revisionist Eduard Bernstein were still alive, he would rub his hands together with glee at some of the public statements Berlinguer and Carrillo have made. There has been no lack of critics to denounce the policy of the Eurocommunists as a decoy, a maneuver of the same sort as the Popular Front and the "outstretched hand" of Stalin's day. There is no question that opportunistic tactics play a fairly large role in the positions taken by the Eurocommunists. There is also something else involved, however, something of greater importance. Eurocommunism has been an attempt on the part of its leaders to respond to the social and historical changes that have taken place on the Continent during the last thirty years. It is a moment in a long and tortuous process of revision and criticism that began long ago and has not ended.

The origins of this process lie in the controversies and polemics that successively tore apart the First, the Second, and the Third Internationals. What is being argued about today had already led, more than a century ago, to fierce disputation, although in another language and from other perspectives, between Marx and Bakunin, between Martov and Lenin—that is to say, among all the leaders of the workers' movement. In the contemporary era the process of revision and criticism was unleashed by the Khrushchev Report. For years the leaders of the communist parties had kept Soviet realities hidden: institutional terror, the servitude of workers and peasants, the regime of privileges, the concentration camps, and, in a word, all of those practices that communists chastely call "violations of socialist legality." Through a moral and psychological mechanism that has yet to be described, Thorez, Togliatti, La Pasionaria, and the others not only accepted the great lie but collaborated actively in perpetuating it. The worst of it was that they managed to preserve the myth of the Soviet Union as the "homeland of the proletariat," not only in the minds of militants but also in the minds of millions of sympathizers. The spectacle of the unshakable faith of innumerable "progressivist intellectuals"—precisely those whose only profession of faith ought to be criticism, examination, and doubt!—was no less scandalous.

But after the Khrushchev Report it was no longer possible to keep the truth under wraps.

In the beginning, the criticism tended to be along moral lines, and this was followed by historical, political, and economic criticism. The task of demolishing an edifice of lies that has stood for more than half a century has not yet come to an end.

As always happens, it was intellectuals—among them many communists—who began the critical examination. It is plain that without the action of the leftist intellectuals the evolution of the European communist parties would have been impossible. Thanks to them, the lies of some ten or fifteen years ago cannot be repeated with impunity today. (This attitude stands in sharp contrast to that of so many Latin American intellectuals, who do not open their mouths except to recite catechisms drafted in Havana.) The criticism of the European intellectuals was effective—unlike the reception given such earlier critics as Serge, Ciliga, Souvarine, Breton, Camus, Silone, Howe, and others—because, almost at the same time, the world discovered the existence of a dissident movement in the Soviet Union. This movement is an oddity in that it is not restricted to a single current: the plurality of tendencies and philosophies that characterized pre-Bolshevik Russia reappears among the dissidents. What is most significant is that Marxists are a minority within the movement.

Another circumstance that accelerated the evolution of the European communist parties was the Russian occupation of Czechoslovakia, the occupation of Afghanistan, and the humiliation of Poland. These were sudden moves that, after the bloody intervention in Hungary, were well nigh intolerable to the European left. The Sino-Soviet conflict was proof that "proletarian internationalism" is the mask of nationalist aggressions; the invasion of Czechoslovakia and the repression in Poland confirm that the interests of the Russian State do not coincide with the interests of the working class or with socialism. The European working class and its leftist intellectuals have taken Russian foreign policy as a double offense: to their socialist feelings and to their nationalist feelings.

The leaders of the communist parties have tried their best to adapt their ideology and their tactics to the realities of the new Europe. It is the Italians and the Spaniards who have gone furthest. In the French party the Stalinist heritage weighs heavily in the balance and pro-Sovietism continues to be *de rigueur*. Why have French socialists, who can look back on many a bitter experience of Communist about-face, decided to govern as "common partners" with them? A holdover of Jacobinism or a Machiavellian maneuver to immobilize them? If the former, it is most regrettable. If the latter, it is a naïve trap that will ensnare no one but the socialists themselves. If circumstances so require, Georges Marchais and the French party will not hesitate to break off the alliance as they have before.[2] Communists look upon ideological tendencies akin to theirs with rage: socialists of every shade, anarchists, labor parties. Not only have they always attacked them, but whenever they have been in a position to do so they have actually persecuted and exterminated them. This characteristic, and the propensity to divide itself and subdivide itself into sects and factions, are proof that communism is not really a political party but a religious order animated by an exclusivist orthodoxy. In the eyes of communists, other parties do not exist, save as groups of individuals who must be either converted or eliminated. For communists, alliance means annexation, and anyone who keeps his independence becomes a heretic and an enemy. Italian communists, it is true, speak of a "historic compromise," a term that implies alliance not only with the other labor parties but with the middle class as well, bourgeois liberals included.

We may nonetheless allow ourselves to wonder whether the policy of the Italian Communists would be the same if there existed in Italy a strong socialist party such as that in France or Spain.

The most spectacular reform has been the abandonment of the dogma of the dictatorship of the proletariat. On this point, it is useful to distinguish between dictatorship of the proletariat and dictatorship of the Communist Party. Marx put forward the former as a basic tenet, not the latter. According to the

2. They actually did so in 1984.

original conception of Marx and Engels, during the transition period leading to socialism power was to be in the hands of the various revolutionary workers' parties. But in the "socialist" countries of today, the communist minority exercises, in the name of the proletariat, a total dictatorship over all classes and social groups, including the proletariat itself. The abandonment of the notion of the "dictatorship of the proletariat" has been yet another sign that the European left, not excluding the communists, is finally beginning to retrieve its *other tradition*. Not the one that stems from Rousseau's "general will," the mask of tyranny and the intellectual origin of Jacobinism and Marxism-Leninism, but the libertarian, pluralist, and democratic one, founded on a respect for minorities.

Despite the importance of the changes that have taken place in the communist parties of Italy and Spain, these have not yet fully evolved. The European communist parties—the French one in particular—continue to be closed groups, at once religious and military orders. The truth is that, if the goal is to return to the true socialist tradition, there is a twofold moral and political prerequisite that must be satisfied. It is necessary, first, to break with the myth of a socialist Soviet Union, and, second, to establish internal democracy within communist parties. This latter condition means revising the Leninist tradition at its very roots. If communist parties wish to cease being religious and military orders so that they may become genuine political parties, they must begin by practicing democracy at home and denouncing tyrants wherever they may be, in Chile or in Vietnam, in Cuba or in Iran.[3]

My critique of European communist parties should not be

3. The failure to achieve a thoroughgoing and truly democratic reform of their structures explains the progressive and, in my opinion, irreversible decline of the French and Spanish communist parties. Marxism-Leninism is no longer a European ideology; it has become a political and military catechism of the revolutionary elite of the less developed countries, such as Nicaragua and Ethiopia.

regarded as an attempt to exculpate the other parties. All of them are more interested in reaching power or remaining in power than in shaping the future. No idea of change inspires them, nor do they represent anything new in the history of this century. Their idea of movement is the in-and-out of office holding, the shuffling of Cabinet posts: politics as a game of musical chairs. I am not unaware that the leaders of the liberal democracies have been skillful and efficient, or that they have resolved many conflicts and problems in a civilized manner. Their countries have great material, technical, and intellectual resources to draw on; they have resisted the old imperial temptation and have husbanded these resources carefully. But at the same time they have not known how, or have not wanted, to use their riches and their technical knowledge for the benefit of poor and economically underdeveloped countries. This has been disastrous, since these countries, in Asia and Africa as in the Americas, have been and will be focuses of disturbances and conflicts. But if Western leaders have not been generous as well as prudent, neither have they committed great excesses. None of them has been a bloodthirsty despot, and all of them have endeavored to respect minorities as well as the majority. The leaders' great errors and misdeeds have taken the form, rather, of sexual or financial scandals. They have exercised power—or the risks of opposition—with moderation and relative intelligence.

This picture would be incomplete if I did not add that their politics has been the politics of complacency, of the easy way out. Idolators of the *status quo* and specialists in the art of wheeling and dealing, they have displayed precisely the same weakness of will in the face of the incredible selfishness of the masses and the elites of their countries as in the face of the threats and blackmail of foreign countries. Their vision of history is that of trade, and it is for this reason that they have seen in Islam not a world awakening but a client to be bargained with. Their policy toward Russia—I am thinking not only of social democrats such as Willy Brandt and Helmut Schmidt but also of conservatives such as Valéry Giscard d'Estaing—has been and is a gigantic self-deception. The essential thing has been to wriggle out of difficult situations, assure themselves of another year of tranquil digestion, and win the next election. There is a disproportion—I don't know whether to call it comic or tragic—between this slippered ease and the decisions that the present demands.

Nonetheless, it would not be fair to fail to acknowledge the great benefits that workers and the middle class have attained in the last forty years. These gains are owed, above all, to the labor unions and to the action of social-democratic and labor parties. To these causes must be added, as a basic economic condition, the extraordinary productive capacity of modern industrial societies and, as a no less basic social and political condition, the democracy that has allowed the struggle and the negotiation between capitalists and workers, and between both and governments, to take place. Productive capacity, free labor unions, the right to strike, the power to negotiate: this is what has made the Western democracies viable and prosperous.

How much longer will the governments of the West be able to assure their peoples that they will enjoy this well-being that, though it may be neither happiness nor wisdom, has been and is a sort of placid contentment based on work and consumption? As the economic crisis worsens, there will be less work, fewer things to buy, and less money with which to buy them. The magnitude of the problems with which we of the twentieth century are confronted stands in sharp contrast to the modesty of the programs and solutions put before us by the governments and political parties of Western Europe. There are those who, not without reason, will surely remind me that politics is an art (or a technique) that lives and moves and has its being in the relativity of the immediate and the proximate. The political leaders of antiquity were likewise not able, save in rare instances, to predict the future: they hit upon the right solution because they knew how to answer the challenge of the present, not because they were able to see into the future. That is true, but we live at a crossroads of history, and Europe is the great absentee in world politics.

The decline of European influence cannot be attributed solely to its political leaders' lack of political imagination or boldness. Following World War II, the nations of the Old World withdrew into themselves, and since then have devoted their immense energies to creating a prosperity without grandeur and cultivating a hedonism without passion and without risks. The last great attempt to regain the lost influence was made by General De Gaulle. With him ended a tradition that even in his own day, and despite his powerful personality, was an archaism. It was simply not possible for France, by itself, in its new secondary rank on the world scene, to re-establish the international balance of power and act as a counterweight to the United States and Russia—a role that could be filled only by the combined forces of a united Europe. But this possibility never really played a part in General De Gaulle's vision, for two reasons: first, because he was a profoundly nationalistic political leader, and, second, because his great intelligence was in no way incompatible with a goodly share of realism. De Gaulle knew what we all know: the nations of Europe want to live together and prosper in peace, but they do not want to do anything in common. The one thing that unites them is their passivity in the face of destiny. Hence the fascination exerted on their peoples by pacifism, not as a revolutionary doctrine but as a negative ideology. This is the other face of terrorism: two contrary expressions of the same nihilism.

In the last few years we have witnessed the electoral triumph of democratic socialism in Spain, France, and Greece. These victories bear within them lessons that should be pondered by all Latin Americans, especially those who are democrats and socialists. The case of Spain is especially pertinent. Spaniards and we Hispano-Americans have confronted the same obstacles in our attempts to implant democratic institutions in our lands. I shall not deal at length with this subject: it is a familiar theme, and I, too, have discussed it in a number of my writings. The history of Spain and of Hispano-America over the last two centuries has given me occasion to doubt more than once whether democracy is viable in our case. Spain today, after forty years of dictatorship, is beginning to live a democratic life that is in many respects exemplary. The first lesson, above all for our obtuse, barbarian oligarchies ever in search of an imposing brass hat to guarantee order: if Spaniards have managed to live together peacefully and democratically, why can't we do likewise? The second lesson is of particular concern to the Latin American left, dogmatic and stubborn-minded, descended not from the Enlightenment but from the theologians of the sixteenth century: the Spanish Socialist Workers' Party (PSOE) has not merely renounced Marxism but has willingly gone along with the democratic rotation of parties and power. Perhaps in view of the example of Spain, our political leaders, both conservative and radical, will learn to practice tolerance, accept criticism, and respect the opinion of others.

The pragmatism of the democratic parties, especially the social democrats, has positive aspects, virtues that become visible in the light of the critiques of revolutionaries. If we reread the polemic between Karl Kautsky and the Bolsheviks today, we are most likely to side with Kautsky: his position vis-à-vis the communist dictatorship is not very different from that of Enrico Berlinguer and Santiago Carrillo today. Yet the name of the German Marxist has been coupled for half a century now with Lenin's damning epithet: Kautsky the Renegade. His case is akin to that of Julian, the Roman emperor in the tradition of Marcus Aurelius (and a valiant warrior) who through the work of his Christian enemies is known today as Julian the Apostate. There is no doubt that socialists and social democrats have ceased to be revolutionaries. Have they not thereby proved that they possess a more acute historical sensibility than their dogmatic critics? The absence of proletarian revolutions in Europe has proved the central prophecy of Marxism to be false. But are the European communist parties of today revolutionary? Abandoning revolutionary rhetoric is a sign not only of intellectual sobriety but also of political honesty.

Since the last years of the nineteenth century, we have lived the myth of Revolution, as the men of the first Christian generations lived the myth of the End of the World and the imminent Second Coming of Christ. I confess that, as the years go by, I view revolt with more sympathy than I do revolution. The former is a spontaneous and almost invariably legitimate uprising against an unjust power. The cult of revolution is one of the expressions of modern excess—an excess that at bottom is an act of compensation for an inner weakness and a lack. We seek from revolution what our elders sought from religions: salvation, paradise. Our era turned gods and angels out of heaven, but it inherited from Christianity the age-old promise to change mankind. From the eighteenth century on, it was thought that this change would consist of a superhuman, though not a supernatural, task: the revolutionary transformation of society. This transformation, like the grace of old, would make men and women *other*. The failure of the revolutions of the twentieth century has been tremendous and is quite evident. Perhaps the Modern Age has been guilty of a terrible confusion: it has tried to make of politics a universal science. It has been believed that revolution, once it became a universal science, would be the key to history, the open-sesame that would open the doors of the prison in which humanity has lived since its very beginnings. We now know that this key has not unlocked a single prison door: it has instead locked many.

The conversion of revolutionary politics into a universal science capable of changing humanity was a transformation of an essentially religious type. But politics is not, nor can it be, anything other than a practice and, at times, an art: its sphere is immediate, contingent reality. Nor has science (or, more exactly, the sciences) ever proposed to change humanity, only to know it and, if such a thing can be possible, to cure it, to better it. Neither politics nor the sciences can bring us paradise or eternal harmony. Hence to turn revolutionary politics into a universal science was to pervert both politics and science, to make of them a caricature of religion. Today we are paying for this confusion in blood. The pragmatism of social democracy, its gradual loss of the radicalism and the vision of justice that had inspired it in the beginning, can be seen as a reaction against the excesses and the crimes of authoritarian and dogmatic socialism. This reaction has been salutary; at the same time, social democracy has been unable to fill the vacuum left by the failure of the great communist hope. Does this mean, as many predict, that the hour of the churches has come? If this should turn out to be the case, I hope that there will be left on the earth at least a small handful of human beings—as at the end of antiquity—who will resist the temptation of divine omniscience, as others, in our day, have resisted that of revolutionary omniscience.

II

Imperial Democracy[1]

The Discovery of Decadence

At first it was a secret passed on in whispers by a handful of those who had caught on; then the experts began to write learned essays in professional journals and give university lectures; today it is a favorite subject of round-table discussions on TV, in newspaper articles, surveys, and popular seminars, in bars, at cocktail parties and dinners. In less than a year Americans discovered that they had "become decadent." Like the divinity of theologians, decadence is indefinable; like springtime in Antonio Machado's famous poem, it comes from nobody knows where; and like both, it is everywhere. Some have greeted the news with shocked disbelief, others with annoyance, and still others with indifference. The religious-minded view it as a punishment from heaven, and deep-dyed pragmatists as a reparable mechanical breakdown. Most have received it with a sort of ambiguous frenzy, a peculiar combination of horror, excitement, and a curious sense of relief: "At last!"

---

1. These pages were written in 1980, and since then there have been many changes in the United States. Nevertheless, I have not revised anything; the changes do not invalidate my observations—indeed, in many cases they confirm them.

From the very birth of their nation, Americans have been a people rushing toward the future. The entire prodigious historical career of the United States can be viewed as a never-ending gallop toward a promised land: the kingdom (or, rather, the republic) of the future. A land made not of solid earth but of an evanescent substance: time. The moment one reaches out to touch the future, it disappears, only to appear again a moment later, just a little bit farther on. Always a little farther on. Progress is as elusive as a ghost. But today, just as Americans were beginning, literally, to run out of breath, the future has suddenly descended in a form at once abominable and infinitely seductive: decadence. The future has a face at last.

The allurements of decadence, though less well publicized than those of progress, are more urbane, subtle, and philosophical: doubt, pleasure, melancholy, despair, remembrance, nostalgia. Progress is brutal and insensitive, has no notion of nuance or of irony, speaks in proclamations and watchwords, is forever in a hurry, and is brought up short only if it crashes headlong into a wall. Decadence commingles the sigh and the smile, the cry of pain and the moan of pleasure, weighs each instant and dallies even in the midst of cataclysm: it is an art of dying, or, better put, of living while dying. I nonetheless believe that the fascination that decadence holds for Americans lies not so much in its philosophical and aesthetic charms as in the fact that it is the gateway to history. Decadence affords them what they have always sought: historical legitimacy. Religions jealously guard the keys to eternity, which is the negation—or, rather, the dissolution—of history. Decadence, on the other hand, makes available to parvenu peoples—be they Romans or Aztecs, Assyrians or Mongols—that modest substitute for eternal glory, earthly fame. Americans have felt their radical modernity to be an original historical sin. Decadence washes away this stain.

For all civilizations, barbarians have been, invariably, men "outside of history." This condition of being "outside of history" has always referred to the past: barbarism is pure ante-riority, the original condition of men before history. Through a singular inversion of the usual perspective, American modernity, the consequence of four thousand years of European and world history, has been regarded as a new barbarism. The civilized person sees overemphasis on the past or on the future as two parallel though antagonistic forms of the eccentricity of barbarians. To the European mind, the future of North Americans was no less uninhabitable than the past of primitive peoples. This sentiment was shared by a number of distinguished Americans who might be called "fugitives from the future": Henry James, George Santayana, T. S. Eliot, and others.

As Europeans were not able to recognize themselves in nomad societies—which were the irrevocable past—they were likewise unable to recognize themselves in American modernity. The United States as a country without Romanesque churches or Gothic cathedrals, without Renaissance painting or Baroque fountains, without a hereditary nobility or an absolute monarchy. A country without ruins. What is most surprising is that Americans, with a few rare exceptions, accepted this verdict: a people "outside of history" was a barbarous people. Hence they endeavored by every possible means to justify their anomaly. The justification took many forms. In literature it went by the name of Melville, Emerson, Whitman, Twain. Today, thanks to the unexpected appearance of decadence, the historical anomaly has ended and the United States has entered into normality. It can unblushingly identify itself with the great empires of the past. It has regained mortality: it now has a history.

The Americans have not been left to exult in their recently discovered decadence undisturbed: the envy of Europeans, the resentment of Latin Americans, and the rancor of the other peoples accompany them. These sentiments, too, are historical—that is to say, they fulfill the same function as the idea of decadence. They are at once a compensation for and a proof of the existence of a great empire, an inverse form of admiration. They thus serve as evidence of a unique, a singular history. Quevedo, who lived amid decadence and hence was a great expert on forms of envy and rancor, places in the mouth of Scipio Africanus, the general who defeated the Carthaginians but was defeated by his fellow Romans, these arrogant words:

> *Nadie llore mi ruina ni mi estrago*
> *pues será a mis cenizias, cuando muera,*
> *epitafio Anibal, urna Cartago.*[2]

Quevedo's sonnet gives us a clue to another possible meaning of the current popularity of the subject of decadence. Scipio recognizes that he was defeated not by the enemies of Rome but by his political rivals. His fate was that of so many heroes who have been brought to ruin in democratic republics, those great breeding grounds of envy and demagogism. I wonder whether the subject of American decadence may not be closely linked to the present electoral campaign.[3] It is an argument to win votes, a projectile launched against the rival candidate. Although (or perhaps because) this campaign has so far lacked grandeur, it is a good example of the endemic disease of democracies: internal dissension, the battle of factions.

Again and again we have seen Americans—intellectuals and journalists in particular, but also former officials—criticize the foreign policy of their country, almost invariably on partisan grounds, because it is the policy of an administration of the opposing party.

Needless to say, I applaud the attitude of Americans: criticism is the very basis of true freedom. Yet at the same time I

2. "Let no one mourn my ruin or my destruction
   since for my ashes, when I die,
   Hannibal will be the epitaph, Carthage the urn." (TRANS.)
3. These lines were written during the 1980 Carter-Reagan campaign.

deplore the way in which they express their criticisms, not because they ought to tone them down but because they need to prevent this exercise of their freedom from being used by the very enemies of freedom. Almost always, such criticisms are picked up by Russian propagandists and spread all over the globe. This attitude of Americans, moreover, is yet another example of their insensitivity to the outside world: they are, truly, outside of history. It goes without saying that domestic disagreements and the electoral campaign do not entirely account for this appearance of the theme of decadence, which, it is evident, is not a fiction for propaganda purposes but a reality. But this reality has been exaggerated—or, rather, distorted.

I am mistrustful of the word "decadence." Verlaine and Montezuma, Louis XV and Góngora, Boabdil and Gustave Moreau have been called decadent for different and contrary reasons. Montesquieu and Gibbon, Vico and Nietzsche have written admirable pages on decadent empires and civilizations. Marx prophesied the end of the capitalist system, Spengler diagnosed the twilight of the culture of the West, Benda analyzed that of "Byzantine France" . . . and so on. To which of all these decadences are we referring when we speak of the United States in 1980?

Despite these uncertainties and vaguenesses, almost all of us share the idea—or, rather, the feeling—that we are living in a twilight era. But the term "decadence" is no more than a rough description of our situation. We are not faced with the end of an empire, a civilization, or a system of production. The evil is universal; it is corroding all systems and infesting all continents. The theme of the general crisis of civilization is not new: for more than a hundred years, philosophers and historians have written books and essays on the decline of our world. The twin of this theme, on the other hand—that of the end of this world—has always been the domain of religious thought. It is a belief that many eras and peoples have shared: the Hindus, the Sumerians, the Aztecs, the first Christians and those

of the year 1000. Today the two themes—of historical decadence and of the end of the world—have fused, forming a
single one with overtones by turn scientific and political,
eschatalogical and biological overtones. We are not merely experiencing a crisis of world civilization: it is a crisis that may
culminate in the physical destruction of the human species as
well.

The destruction of the planet Earth is an event that neither
Marx nor Nietzsche nor any other of the philosophers who
have pondered the theme of decadence envisioned in their
writings on the subject. As for religious thought: although its
specialty is the death and birth of humankind and of societies,
religious traditions have always held that the world would be
destroyed by supernatural beings or by cosmic forces, not by
the action of human beings using technological means. Modern science, for its part, has speculated a great deal on the
ultimate end, by which I mean that of the entire universe, not
just of our planet; the Second Law of Thermodynamics—the
progressive loss of heat and the descent into a state of inert
disorder that has no end—has been and is our Trumpet of the
Last Judgment. Yet the ultimate degradation of energy will be
the work not of men but of the very economy of nature. The
ancient philosophers pondered the question of whether the
universe was doomed to extinction. Some inclined toward the
hypothesis of a self-sufficient and eternal universe; others toward the cyclical vision: the conflagration *(ekpyrosis)* that, according to the Stoics, ends a cosmic period and at the same
time kindles the fire of universal resurrection. Modern philosophy has not taken up the theme of the end of the world or
the other cosmological speculations of antiquity. It has, it is
true, reflected on individual death and on the decadence of
societies and civilizations, but the extinction of our planet has
been a subject left to physics and the other natural sciences.

In the second half of the twentieth century, the end of the
world has become a public affair, and one to be dealt with

exclusively in terms of human beings and their acts. It is neither demiurges nor natural forces, but human beings alone who will be responsible for the extinction or the survival of their species. This is the great historical novelty of our century—an absolute novelty, and one that may mean the end of all novelties. If such a thing comes about, destiny will have cured humanity, in a terrible and an absolute way, of the sickness from which it has suffered since its beginning, and which, having flared up again more than two centuries ago, has now reached an acute stage: the greed for novelties, the insane cult of the future. Like the souls in Dante's Inferno, we would be doomed to the *abolition of the future*, except that, unlike them, we would not even be able to witness this unthinkable event. In truth, our fate would be—sinister symmetry—exactly the contrary of theirs: eternal death. Our era would thereby realize itself completely, drain to the dregs its destiny, the negation of Christianity.

The United States is a part—and an essential part—of the general crisis of civilization; it also shares with the Soviet Union the terrible responsibility for creating the situation that may bring the end of the human species, and perhaps of life itself, on this planet. But it is obvious that, when we speak of the decadence of the imperial republic of the United States, we are referring to something quite different. From the perspective of the worldwide crisis of civilization, the United States has suffered less than almost any other nation the horrors and ravages of our era. Although it has experienced many vicissitudes and undergone enormous changes, its political, social, and economic foundations are still intact. American democracy has managed to correct, to a large degree though not entirely, its grave imperfections in the area of the rights of ethnic minorities. In the domain of individual freedoms and respect for privacy and private morals, there has also been visible improvement. And, finally, Americans have not known totalitarianism, as have the Germans, the Russians, and the nations living under Soviet domination. Their country has not been occupied, and they have not seen their cities destroyed; nor have they suffered the dictatorships, civil wars, famines, humiliations, exactions that so many other peoples have.

Faced with the concrete reality of the United States, the first, the natural reaction of any visitor is utter amazement. Few have gone beyond that initial shock of surprise—admiration mingled at times with revulsion—to realize the immense originality of that country. One of those few, and the first of them, was Tocqueville. His reflections are still as pertinent as ever. He foresaw the future greatness of the American Union and the nature of the conflict that has lain at its heart ever since its birth, a conflict to which it owes, at one and the same time, both its great successes and its great setbacks: the opposition between freedom and equality, the individual and democracy, local freedoms and federal centralism. Henry Adams's vision, though less broad, was perhaps more profound: deep within American society he saw an opposition between the Dynamo, which transforms the world but reduces it to uniform series, and the Virgin, the natural and spiritual energy that irrigates and illuminates the human soul and thus produces the range and variety of our works. Tocqueville and Adams saw, clearly and sharply, what was going to happen; we, today, see what is happening. In this historical perspective my reflections are perhaps not altogether idle.

When I speak of originality, I am not referring to the contrasts with which we are all familiar—great wealth and extreme privation, the cheapest vulgarity and the purest beauty, greed and altruism, active pursuit of goals and the passivity of the drug addict or the frenetic violence of the drunkard, proud freedom and the docility of the herd, intellectual exactitude and the fuzzy delirium of the nut case, prudishness and license—but, rather, to the *historical novelty* that the United States represents. Nothing in our human past has been comparable to this reality that is made up of violent clashes and glaring contrasts, and is, if I may use the expression, full of itself. Full and empty: what lies behind this tremendous variety of products and goods flaunted before the eyes of the world with a sort of shamelessness born of generosity? A wealth that is fascinating—that is to say, deceptive. I am not thinking of the injustices and inequalities of American society: though they are many, they are fewer and less grave than our own, than those of most nations. I say "deceptive" wealth not because it is unreal but because I ask myself whether a society can live trapped within the confines of the circle of production and consumption, work and pleasure. There are those who will say that this situation is not unique, but common, rather, to all the industrial countries. That is true, but in the United States, since it is the nation that has gone the farthest along this path and is thus the most perfect expression of modernity, the situation has reached its extreme limit. What is more, within this situation there is a most unusual note sounded that we do not hear from other nations.

I repeat my question: what lies behind this wealth? I cannot answer; I find 'nothing, there is nothing. I explain myself: all institutions in America—its technology, its science, its energy, its education—are a means, a *way toward*. Freedom, democracy, work, inventive genius, perseverance, fulfillment of promises and obligations: everything is *useful*, everything a means to attain—what? Happiness in this life, salvation in the life beyond, the good, truth, wisdom, love? Ultimate ends, those that really count because they give meaning to our lives, are not visible on the horizon of the United States. They exist, that is certain, but they appertain to the private domain. Questions and answers as to life and its meaning, death and the life beyond, traditionally taken over by Church and State, have heretofore always been matters in the public domain. The great historical novelty of the United States lies in its attempt to return them to the private domain, the private life of each and every citizen. What the Protestant Reformation achieved in the

sphere of beliefs, the American Union has achieved in the secular sphere. A tremendous novelty, a change without precedent: what, then, is left of the action of the State—that is to
say, of history?

American society, unlike all other societies we know of, was
founded in order that its citizens might realize their private
ends in peace and freedom, on the theory that the common
good lies not in a collective or metahistorical end but in the
harmonious coexistence of individual ends. Can nations live
without common beliefs and without a metahistorical ideology? In the past, the acts and deeds of each people were nourished and justified by a metahistory; in other words, by a
common end that lay above individuals and had to do with
values that were, or were presumed to be, transcendent.
Americans naturally share beliefs, values, and ideas: freedom,
democracy, justice, work, and so on. But all such concepts are
a means, something *for* this or that. The ultimate ends of their
acts and thoughts lie not in the public domain but in the private. The American Union was the very first historical attempt
to give back to the individual what the State had stolen from
the person in the beginning.

I do not mean by that that the American State is the only
liberal State: its founding was inspired by the examples of Hol-
land, England, and the philosophy of the eighteenth century.
But the American nation, and not only the State, is different
from others precisely because it was founded on these ideas
and principles. Unlike what happened elsewhere, the United
States Constitution does not modify or change a prior situation
(in its case, the monarchical regime with its hereditary classes,
estates, and special jurisdictions); it institutes, rather, a new
society. It marks an absolute beginning. It has frequently been
said that in liberal democratic societies, especially in the United
States, the power of individuals and groups, above all of capitalist enterprises but also of workers' bureaucracies and other
sectors, has grown without restraint, to the point where State

domination has been replaced by that of special interests. The criticism is a fair one. It must be added, however, that while this reality seriously distorts the original design, it does not nullify it altogether. The founding principle is still alive. Proof of that can be found in the fact that it continues to inspire the movements of self-criticism and reform that periodically shake the United States. All of these have represented themselves as a return to the country's origins.

The great historical originality of the American nation, and at the same time the root of its contradiction, lies in the very act by which it was founded. The United States was founded in order that its citizens might live, among themselves and by themselves, free at last of the weight of history and the meta-historical ends that the State assigned societies in the past. It was a construct aimed against history and its disasters, oriented toward the future, that *terra incognita* with which it has identified itself. The cult of the future fits naturally within the American design and is, so to speak, its condition and its result. American society was founded by an act of abolition of the past. Its citizens, unlike Englishmen or Japanese, Germans or Chinese, Mexicans or Portuguese, are not the offspring of but the beginning of a tradition. Instead of carrying on a past, they inaugurate a new time. The act (and the document) of foundation—a canceling-out of the past and a beginning of something different—has constantly been repeated throughout its history: each one of its episodes is given its particular configuration with reference not to the past but to the future. It is a step forward. Toward where? Toward a nowhere that is everywhere—except here and now. The future has no face and is sheer possibility. . . .

But the United States is not in the future, a region that does not exist; it is here and now, among all the rest of us, in the midst of history. It is an empire, and its slightest movements shake the whole world. It would like to be outside the world but it is in the world—it is the world. Hence the contradiction of contemporary American society: being at once an empire and a democracy is the result of another, deeper contradiction, having been founded against history yet being itself history.

On a recent trip to the United States I was surprised by the plethora—in the windows and on the shelves of bookstores in New York and Cambridge—of books and periodicals dealing with the subject of decadence. The publications satisfy, on the one hand, the American tendency toward self-criticism and self-flagellation; on the other, they are fabrications of the publicity industry. In a society ruled by the cult of fashion—a cult that is also a business—even the theme of decadence becomes both a novelty and a commercial proposition. Many of these books and articles on the twilight of the United States are, in the twofold sense of the word, speculations. At the same time, willy-nilly, they fulfill a psychological and moral function that I am uncertain whether to call a compensation or a purification. Americans today devote themselves with a sort of single-minded greed to the arduous pleasures of scrutinizing their consciences. A sign of morbid inclinations or a search for salvation?

Amid the flood of books, essays, and articles on the decline of the United States, a distinction must be made. Many of these are elaborate exercises of the imagination, more or less intelligent variations on one of those collective fantasies that our world periodically secretes in its continual thirst for novelties and cataclysms. Others, more serious, are concrete analyses of specific problems, and of carefully defined areas: military questions, international relations, economic affairs. Almost all of these are persuasive, and after reading them, it is difficult not to agree that for years now we have been witnessing a gradual weakening of the military and political strength of the United States. The imperial republic, as many signs warn us, has already reached its zenith, and probably its decline has already begun. This is a slow process and may go on for one century, as in the case of the Spanish Empire, or five, as in that of

Rome. Except that, contrary to what has happened in the past, a new star is yet to be seen rising on the historical horizon. The ills and contradictions of its great rival are graver still, and perhaps incurable. The Soviet Union is a society of castes and a multinational empire under the domination of Great Russia. It thus lives caught between the threats of petrification and explosion.

Epicurus or Calvin?

The United States is undergoing a period of doubt and disorientation. If it has not lost faith in its institutions—Watergate was an admirable proof of this—it no longer believes as fervently as it once did in the destiny of the nation. Within the limits of the present essay, it is impossible to examine the reasons and the causes for this: they belong to the realm of the "long count." Suffice it to say that the present state of mind of the American people is in all likelihood the consequence of two phenomena that used to be opposites but, as frequently happens in history, have now become conjoined. The first is the sense of guilt that the Vietnam War aroused in many minds; the second is the waning of the puritan ethic and the waxing of the hedonism of abundance. The sense of guilt, coupled with the humiliation of defeat, has reinforced the traditional isolationism, which has always regarded American democracy as an island of virtue in the sea of perversities that is world history. Hedonism, for its part, takes no notice of the outside world or, along with it, of history. Isolationism and hedonism coincide in one respect: they are both antihistorical. Both are expressions of a conflict present in American society since the war with Mexico in 1847, but not fully apparent until this century: the United States is a democracy and at the same time an empire. A peculiar empire, I must add, for it does not wholly fit the classic definition of one. It is something quite distinct from the Roman, Spanish, Portuguese, and British empires.

Standing bewildered in the face of its dual historical nature, the United States does not know which way to turn today. The dilemma is a fateful one. If it chooses an imperial destiny, it will cease to be a democracy and will thereby lose its reason for being as a nation. But how to renounce power without being immediately destroyed by its rival, the Russian Empire? It will be objected that Great Britain, too, was both a democracy and an empire. The contemporary situation is very different, however: Great Britain's imperial rule was exclusively colonial and exercised overseas; moreover, in its European and American policy it sought not hegemony but a balance of power. But the policy of the balance of powers belongs to another stage in world history; neither Great Britain nor any of the other great European powers was forced to confront a State such as the Soviet Union, whose imperialist expansion is inextricably linked to a universal orthodoxy. The Russian Bureaucratic State not only aspires to world domination but is a militant orthodoxy that does not tolerate other ideologies or other systems of government.

If, instead of comparing the international situation that confronts the United States today with that which prevailed in Europe during the second half of the last century, we think of Rome in the last days of the Republic, the comparison shows American democracy to be in an even more unfavorable position. The political difficulties of the Romans of the first century B.C. were primarily internal in nature, and this partially explains the ferocity of the struggles among the various factions: Rome had already achieved domination over all the known world, and its only rival—the Parthian Empire—was a power on the defensive. Moreover, and most important: none of the powers that had fought the Romans sought to further a universalist ideology. By contrast, the contradictions of American foreign policy—a result of the controversies among groups and

parties as well as of the inability of the nation's leaders to formulate a long-term overall plan—coincide with the existence
of an aggressive empire that embraces a universalist ideology.
To make matters still worse, the Western alliance is made up
of countries whose interests and politics are not always identical with those of the United States.

The expansion of the American republic has been the natural, and in some ways fatal, consequence—if I may so put it—
of its economic and social development; Roman expansion grew
out of the deliberate action of the senatorial oligarchy and its
generals over a period of more than two centuries. The foreign
policy of Rome is an outstanding example of coherence, sin-
gleness of purpose, perseverance, skill, tenacity, and prudence—precisely the virtues that we find lacking in Americans.
Tocqueville was the first to see where the fault lay:

With regard to the conduct of the external affairs of society,
democratic governments appear to me to be decidedly inferior
to the others. . . . Foreign policy requires the use of almost
none of the qualities that characterize democracy, and on the
other hand calls for the development of almost all those which
democracy lacks by its very nature. . . . Democracy would
find it most difficult to coordinate all the details of a great undertaking, draw up a plan in advance, and stubbornly follow
it to the end despite all obstacles. It has little aptitude for preparing its means in secret and patiently awaiting the results.
These are qualities appertaining to a man or to an aristocracy;
and these are precisely the qualities that in the long run enable
peoples, acting as one individual, to dominate others.

American democracy is religious in origin and can be traced
back to the communities of Protestant dissenters who settled
in the country during the seventeenth century. Religious
preoccupations were later transformed into political ideas steeped
in republicanism, democracy, and individualism, but the orig-

inal religious tone never disappeared from the public conscience. In the United States, religion, morality, and politics
have been inseparable. This is the major difference between
European liberalism, which is almost always secular and anti-
clerical, and the American variety. Among Americans, democratic ideas have a religious foundation, in some instances
implicit and in others (the majority) explicit. These ideas served
to justify the attempt, unique in history, to constitute a nation
as a *covenant* in the face of, and even against, historical necessity or history as fate. In the United States the social contract
was not a fiction but a reality, and it was entered into in order
not to repeat European history. This is the origin of American
isolationism: the attempt to establish a society that would escape the vicissitudes that European peoples had suffered. It was
and is, as I have already said, a construct against, or, rather,
outside of history. It follows from this that American expansion, up until the war with Mexico, was aimed at colonizing
empty spaces (Indian peoples were always regarded as *nature*)
and that space more empty still, the future.

If they could, Americans would lock themselves up inside
their country and turn their backs on the whole world, except
to trade with it and visit it. The American utopia—in which,
as in all utopias, monstrous features abound—is an interweav-
ing of three dreams: those of the ascetic, the merchant, and
the explorer. Three individualists. Hence three American traits:
their reluctance to confront the outside world; their inability
to understand it; and their lack of skill in manipulating it.
Americans are citizens of an empire, surrounded by some nations that are its allies and by others that are out to destroy it,
yet Americans would far rather be left to themselves: the outside world is evil, history is perdition. America is the opposite
of Russia, another religious country but one that identifies religion with the Church and finds the confusion between ideology and party legitimate. The communist State—as was quite
evident during the last war—is not only the successor of the

czarist State but its continuer. The notion of a social contract or "covenant" has never held an important place within the political history of Russia, or within the czarist or Bolshevik tradition. Nor has the idea of religion as something belonging to the sphere of heartfelt individual belief; to the Russians, religion and politics appertain not to the sphere of private conscience but to the public sphere. Americans have endeavored, and are endeavoring, to construct a world of their very own, a world outside of this world; the Russians have endeavored, and are endeavoring, to dominate this world in order to convert it.

The basic contradiction of the United States has an effect on the very foundations of the nation. Hence our reflections on the United States and its present predicament lead to the question: will it be able to resolve the contradiction between empire and democracy? At stake are its life and its identity.

Though it is impossible to answer this question, it is possible to venture a comment.

The sense of guilt can be transformed, can lead directly to the beginnings of political salvation. Hedonism, on the other hand, leads only to surrender, ruin, defeat. It is, admittedly, true that after Vietnam and Watergate we have been witness to a sort of masochistic orgy and seen many intellectuals, clergymen, and journalists rend their garments and beat their breasts as signs of contrition. These self-accusations, as a general rule, were not and are not false, but their tone was and is frequently hysterical (as when a journalist, writing in *The New York Times*, held American policy in Indochina responsible for the subsequent atrocities committed by the Khmer Rouge and the Vietnamese). Yet this sense of guilt, besides being a compensation that maintains a psychic equilibrium, carries moral weight: it stems from a searching of conscience and the recognition that a wrong has been committed. Hence it can become a sense of responsibility, the one and only antidote against the intoxication of *hubris*, for individuals as for empires. On the other hand, it is more difficult to transform the skin-deep hedonism of modern masses into a moral force. It is not blind illusion, however, to place our trust in the ethical and religious foundations of America: they are a living source whose flow has been obstructed but not yet entirely dammed.

The foreign policy of the United States has followed a zigzag, erratic course, frequently contradictory and at times beyond all understanding. Its principal defect, its basic inconsistency, is attributable not to the failings of American leaders, which are many, but to its being a policy more sensitive to domestic reactions than to foreign ones. The United States' objectives are to contain the Soviet Union and its shock troops (Cuba, Vietnam), to consolidate its own alliance with Japan and the European democracies, to consolidate its ties with China, to bring about an agreement in the Middle East that will preserve the independence of Israel and at the same time strengthen its friendship with Egypt, to gain friends in the Arab countries and in those of Latin America, Africa, and Asia. These are its avowed ends, but its real ones are to win the votes and satisfy the aspirations and ambitions of this or that group at home, whether Jews or blacks, industrial workers or farmers, the "establishment" of the East or Texans. It is evident that the policy of a great power cannot be subordinated to the shifting and divergent pressures of various groups within the nation: the cause of the downfall of Athens was not so much Spartan arms as the struggles between internal parties.

Any list of the errors of American policy must end with the following reservation: these errors, magnified by the mass media and by political passions, are revealing of vices and faults inherent in plutocratic democracies, but they do not indicate an intrinsic weakness. The United States has suffered defeats and setbacks, but its economic, scientific, and technological power is still superior to that of the Soviet Union. So is its political and social system. American institutions were designed for a society in perpetual motion, whereas Soviet institutions correspond to a static caste society. Hence any change in the Soviet Union endangers the very foundations of the regime. Russian institutions would be incapable of withstanding the test of electing a president by popular vote every four years, as the United States does. A phenomenon such as Watergate would have unleashed a revolution in Russia. There is much talk of the inferiority of the Americans in the military sphere, especially in the area of traditional weapons. This is a temporary inferiority. The United States has the material and human resources to re-establish the military balance of power.

And the political will? It is difficult to give an unequivocal answer to that question. In recent years, Americans have suffered from a psychic instability that has taken them from one extreme to another. Not only have they lost their sense of direction; they have also lost control of themselves. What the United States has lacked is not power but wisdom. Quite apart from the exaggerations of publicity, the word "decadence" is applicable to the United States in a moral and political sense. There is a notable disparity between its power and its foreign policy, between its national virtues and its international acts. . The American people and its leaders lack that sixth sense that almost all great nations have had: *prudence.* Since Aristotle, this word designates the highest political virtue. Prudence is made up of wisdom and integrity, boldness and moderation, discernment and persistence in undertakings. The best and most succinct definition of *prudentia* was given recently by Cornelio Castoriadis: the ability to find one's bearings in history. This is the ability that many of us find lacking in the United States.

The United States is frequently compared with Rome. Though the parallel is not altogether exact—in Rome the utopian ingredient, central to the United States, does not appear—it is admittedly a useful one. To Montesquieu, the decadence of the Romans had a twofold cause: the power of the army and the corruption of luxury. The first was the origin of the empire, the second its ruin. The army gave Rome dominion over the

world but, along with it, irresponsible sybaritism and extravagance. Will the Americans be wiser and more temperate than
the Romans; will they show greater moral fortitude? It seems
most unlikely. However, there is one aspect of the situation
that would have raised Montesquieu's spirits: the Americans
have succeeded in defending their democratic institutions and
have even broadened and perfected them. In Rome, the army
backed the despotism of the Caesars; the United States suffers
from the ills and vices of freedom, not those of tyranny. Though
deformed, the moral tradition of criticism that has accompanied the nation all through its history is still alive. The very
accesses of masochism that overcome the country are sick
expressions of this moral requisite.

In the past, the United States was able to use self-criticism
to resolve other conflicts. It continues to give proof of its ca-
pacities for self-renewal. In the last twenty years it has taken
great strides in the direction of resolving the other great contradiction that tears it apart, the racial question. It is not beyond the bounds of possibility that by the end of this century
the United States will have become the first multiracial democracy in history. Despite its grave imperfections and its vices,
the American democratic system bears out the opinion of antiquity: if democracy is not the ideal government, it is the least
bad.

One of the great achievements of the American people has
been to preserve democracy in the face of the two great threats
of our day: the powerful capitalist oligarchies and the bureaucratic State of the twentieth century. Another positive sign:
Americans have made great advances in the art of human co-
habitation, not only in terms of different ethnic groups that
live peacefully together but also in domains heretofore ruled
by the taboos of traditional morality, such as sexuality. Some
critics lament the permissiveness and the relaxation of morals
of American society; I confess that the other extreme strikes
me as worse—the cruel puritanism of communists and the

bloody prudery of Khomeini. Finally, the development of the sciences and technology is a direct consequence of the freedom of investigation and criticism predominant in the universities and cultural institutions of the United States. American superiority in these fields is no accident.

How and why, in a democracy that has proved itself to be so endlessly fertile and creative in science, technology, and the arts, should its politics be so overwhelmingly mediocre? Can the critics of democracy be right? We must grant that the will of the majority is not a synonym for wisdom: the Germans voted for Hitler, and Chamberlain was elected democratically. The democratic system is exposed to the same risk as hereditary monarchy; the popular will is no more unerring than the genes, and elections that turn out badly are as unpredictable as the birth of defective royal heirs. The remedy lies in the system of checks and balances: the independence of judicial and legislative power, the weight of public opinion in governmental decisions through the healthy and sensible exercise of their critical function by the communications media. Unfortunately, neither the Senate nor the media nor public opinion has given signs of political *prudence* in the years just past.

The inconsistencies in American foreign policy are attributable not just to officeholders and politicians but to the entire nation. Not only do the interests of groups and parties come before collective ends, but American opinion has shown itself incapable of understanding what is happening beyond its borders. This criticism is as applicable to liberals as to conservatives, to clergymen as to labor leaders. There is no country better informed than the United States; its journalists are excellent and they are everywhere, its experts and specialists have all the data and background facts needed for each case—yet what comes forth from this gigantic mountain of information and news is, almost always, the mouse of the fable. An intellectual failure? No: a failure of historical vision. Because of the very nature of the endeavor that founded the nation—sheltering it from history and its horrors—Americans suffer from a congenital difficulty in understanding the outside world and orienting themselves in its labyrinths.

Another defect of American democracy, already noted by Tocqueville in his day: egalitarian tendencies, which do not suppress individual selfishness but merely deform it. These tendencies have not prevented the birth and spread of social and economic inequalities, while at the same time they have held the best back and hampered their participation in public life. A major example is the situation of the intellectual class: its first-rate achievements in the sciences, technology, the arts, and education stand in sharp contrast to its scant influence in politics. It is true that many intellectuals serve and have served in government, but this has almost always been as technicians and experts—that is, *in order to do* this or that, not in order to help define ends and goals. A few intellectuals have been counselors of presidents and have thus contributed to planning and executing American foreign policy. But they are isolated cases. The American intellectual class, as a social entity, does not have the influence that its counterparts in European and Latin American countries enjoy. For one thing, society is not inclined to grant this class such a role. It is scarcely necessary to recall the derogatory terms used to designate the intellectual: "egghead," "highbrow." These epithets badly damaged the political career of Adlai Stevenson, to cite but one example.

American intellectuals, in turn, have shown little interest in the great philosophical and political abstractions that have roused deep passions in our era. This indifference has had a positive aspect: It has kept them from going as badly astray as many European and Latin American intellectuals. It has also kept them from the despicable moral lapses and relapses of so many writers who, without so much as blinking an eye, have accepted public honors and international prizes as they hymned paeans of praise to the Stalins, the Maos, and the Castros.

Among the great American poets, only one, Ezra Pound, succumbed to totalitarian spellbinding, and it is revealing that he chose to be the panegyrist of the least brutal of the brutal dictators of this century: Mussolini. Unlike certain European and Latin American writers, Pound did not obtain either decorations or national funeral honors in return for his apostasy; he was instead shut up for many years in an insane asylum. This was a terrible thing, but perhaps better than the gleeful splashing about in the mud of an Aragon. The indifference of Americans is not reprehensible in and of itself; it becomes so when it is transformed into the paranoia of conservatives, or the naïveté bordering on complicity of liberals. There are two ways of ignoring the existence of others: by turning them into devils, or by turning them into heroes of fairy tales. American intellectuals' mistrust of ideological passions is understandable; what is not understandable is their ignoring the fact that these passions have moved several generations of European and Latin American intellectuals, among them some of the best and most generous. In order to understand these others and to understand contemporary history as well, it is necessary to understand these passions.

When the subject under discussion is the American character, the word "naïveté" almost invariably crops up. Americans themselves value *innocence* very highly. Naïveté is not a character trait that fits well with the pessimistic introspection of the puritan. Yet the two coexist within the American character. Perhaps introspection allows them to see themselves and discover, within their heart of hearts, the traces of God or of the devil; naïveté, in turn, is their mode of presentation of self to others and their manner of relating to them. Naïveté is an appearance of innocence. Or, rather, it is protective gear. Thus the apparent defenselessness of one who is naïve is a psychological weapon; it preserves that person from the contamination of the other and, by isolating him or her, makes it possible to escape and launch a counterattack. The ingenuousness of

American intellectuals in the face of the great ideological debates of our century has fulfilled that double function. It has kept them from falling into the moral errors and perversions into which certain Europeans and Latin Americans have fallen; and it has permitted them to judge and condemn those who have strayed from virtue—without understanding them. Both American conservatives and liberals have substituted moral judgment for historical vision. Admittedly, it is not possible to have a view of the other, that is to say a vision of history, without moral principles. But a moral perspective cannot replace true historical vision, above all if this moral perspective is that of a provincial puritanism, combined with variable but strong doses of pragmatism, empiricism, and positivism.

In order to understand more clearly the precise nature of this substitution of moralizing for historical vision, I must go back once more to the origin of the United States. In antiquity, private morality was inseparable from public. In the classical philosophers of Greece, Plato and Aristotle, the union between metaphysics, politics, and ethics was an intimate one: the loftiest individual ends—love, friendship, knowledge, and contemplation—were inseparable from the *polis*. The same is true of the great Romans. I need scarcely call to my reader's mind Cicero, Seneca, and, above all others, Marcus Aurelius. In the days of the Greeks and the Romans, however, the separation between ethics and politics (or, as we would say today, between morality and history) had already begun. For many philosophical schools, the Epicureans and the Skeptics in particular, morality tended more and more to become a private affair. But their indifference to public life did not take on those negative forms of political action represented by civil disobedience and passive resistance. Epicurean morality did not lead to a politics. Nor did Skepticism: although Pyrrho would affirm nothing as certain fact, not even his own existence, his doubts did not keep him from obeying the laws and the authorities of the city.

With Christianity, the break between private morality and politics became complete, though its ultimate aim was to convert private morality into a domain that was also collective: that of the Church. During the Reformation, the most profound moral experience, the religious one, became once again an intimate matter: the dialogue of the creature with himself and with his God. The great historical novelty of the United States lies, as I have said earlier, in its having secularized and generalized the intimate relationship of the Christian with God and with his conscience; and second, and perhaps most important, in having inverted the relationship—subordinating the public to the private.

The antecedents of this great change lay in the seventeenth and eighteenth centuries, in thinkers such as Locke and Rousseau, who saw in the original social compact the foundation of society and of the State.[4] But in their writings the idea of the social contract appears *in the face of* an already constituted society; hence it is presented as a critique of an already existing state of affairs. Locke, for example, attempted to refute the doctrine of the divine right of kings; Rousseau conceived of the social compact as an act prior to history, the nature of which history distorts and debases through its introduction of private property and inequality.

In the founding of the United States, these ideas underwent a radical change. The position of the terms was reversed. The social contract was not prior to history; it became, rather, a historical *project*. In other words: it was no longer something that belonged entirely to the past, but a program whose field

4. It is revealing that in Spanish and Hispano-American political thought of the Modern Age, the Hispanic Neo-Thomists, who were the first to see in social consensus the foundation of monarchy itself, have left almost no trace. This blind spot in our historical conscience is yet another example of a well-known fact: the adoption of modernity coincided with the abandonment of our tradition, including ideas that, like those put forth by the Renaissance theologians Francisco Suárez and Francisco de Vitoria, were closer to modern constitutionalism than were Calvinist speculations.

of realization was the future. Likewise, the space in which the contract was realized was not a land with a history but a virgin continent. The birth of the United States was the triumph of a voluntary contract over the fatality of history, of private ends over collective ends, and of the future over the past.

In the past, history was conceived of as a collective action— a *geste*—undertaken in order to realize an end that transcended individuals and society itself. Society related its acts to an end outside itself, and its history found meaning and justification in a metahistory. The depositaries of those ends were the Church and the State. In the Modern Age the nature and the meaning of the action of society have changed. The United States is the most complete and the purest expression of this change, and it is therefore no exaggeration to say that it is the archetype of modernity. The ends of American society do not lie beyond it, nor are they a metahistory: they lie within it and can be defined only in terms of individual conscience.

What are these terms? As I have already stated: essentially and primordially, the individual's relationship with God and with himself; and secondarily, with others, his fellow citizens. In primitive society the self exists only as a fragment of the great social whole; in American society the social whole is a projection of individual consciences and individual wills. This projection is never geometrical: the image it offers us is that of a contradictory reality in perpetual motion. These two key-notes, contradiction and movement, convey the extraordinary vitality of American democracy and its tremendous dynamism. At the same time, they reveal its dangers: contradiction, if excessive, can paralyze a democracy if it is threatened from without; dynamism can degenerate into meaningless frenzied activity. These two dangers loom large at present.

From this broader perspective it is easier to understand the tendency on the part of American intellectuals to substitute moral criteria—or, worse still, pragmatic and purely circumstantial ones—for historical wisdom. Moralism and empiricism are the twin forms that the lack of understanding of history assumes. Both are a reflection of the basic isolationism of the American mentality, which is in turn the natural consequence of the project underlying the founding of the country: building a society safe from the horrors and accidents of world history. Isolationism contains no elements (except negative ones) from which to develop an international policy. This observation applies to both liberals and conservatives—two terms that, as is well known, do not have the same meaning in the United States as they have in Europe and Latin America. The American liberal is a proponent of State intervention in the country's economy, which brings him closer than European and Latin American liberals to social democracy; the American conservative is an enemy of State intervention both in the economy and in education, attitudes that are not very far removed from those of our liberals. With respect to international affairs, however, the positions of liberals and conservatives are interchangeable: both shift quickly from the most passive isolationism to the most determined interventionism, though these shifts do not substantially modify their vision of the outside world. It is not strange, therefore, that despite their differences both liberals and conservatives have been by turn interventionists and isolationists.

My description of the attitudes of American intellectuals is, I grant, quite incomplete. I am not unaware of the existence of currents more closely akin to the tradition of continental Europe and less affected by what it is not going too far to call Anglo-Saxon eccentricity. In the recent past, for example, certain writers strongly influenced by T. S. Eliot—the so-called Fugitives—sought in a mythical South an order of civilization that at bottom was simply a transplant of preindustrial European society. A dream rather than a reality, yet one that broke through the historical solitude of the United States and reunited these nostalgic writers with European history. Moved by a similar impulse, though in the opposite direction, a group of New York intellectuals founded the *Partisan Review* (1934). This periodical went from communism to Trotskyism to a broader, more lively, more modern vision of contemporary reality. But throughout these changes, the *Partisan Review* never lost sight of the primordial relation between history and literature, politics, and morality. Through their concerns and their intellectual style, the editorial staff and contributors to the *Partisan Review* were closer to European writers of that period—I am thinking above all of Camus and of Sartre and Merleau-Ponty—than to their American contemporaries. The same can be said, even today, of other writers and isolated personalities, such as Susan Sontag. Nonetheless, however notable they or their contributions were, none of them belongs to the central tradition.

Some years ago, the philosopher John Rawls published a book entitled *A Theory of Justice* (1971), which knowledgeable readers judged to be an outstanding work. The book is indeed marked by both a surprising rigor and great moral elevation, in the best Kantian tradition: clarity of thought and purity of heart. I cite this work precisely because of these eminent qualities, which make it the best possible example of the American indifference to history. Rawls proposed "to generalize and raise to a higher order of abstraction the traditional theory of the social contract, as formulated by Locke, Rousseau, and Kant." His book contains a number of moving chapters on such subjects as the legitimacy of civil disobedience; envy and equality; justice and equity. Rawls ends with a dual affirmation, of the principle of freedom and the principle of justice: they are inseparable. He has elaborated a moral philosophy founded on the free association of men, yet he concedes that the virtue of justice can manifest itself only in a *well-organized society*. He does not tell us how we may arrive at this or what it consists of. A well-organized society, however, can only be a just society. Apart from the circularity of the argument, I am troubled by the indifference of this author, whose rigor in the realm of concepts and meanings is so notable, to the terrible reality of five thousand years of history.

A Theory of Justice is a book of moral philosophy that omits politics and does not examine the relation between morality and history. It thus lies at the opposite pole from European political thought. To bear out this statement, one need only mention the names of writers as diverse as Max Weber, Croce, Ortega y Gasset, Hannah Arendt, Camus, Sartre, Cioran. All of them *lived* (Cioran is *living* still) the great breach between morality and history; some tried to insert morality into history or to deduce from history the foundations of a possible morality. Even Marxists—Trotsky, Gramsci, Serge—became aware of the great rent and tried to justify it or transcend it in one way or another. The lesson of one of these thinkers, Simone Weil, was particularly precious in that it showed that historical necessity cannot be a substitute for morality, which is based on freedom of conscience; at the same time, through her life and her work, Simone Weil showed us that morality cannot be divorced from history. The great wound of the West has been the separation of morality and history. In the United States this division has taken on two parallel expressions: empiricism on the one hand, and moral abstractions on the other. Neither the one nor the other can effectively combat our modern leprosy: the usurpation, in the communist countries and in many other nations, of morality in the name of a historical pseudo-necessity. The secret of the resurrection of the democracies—and hence of true civilization—lies in the re-establishment of the dialogue between morality and history. This is the task of our generation and the one to come.

The United States was the first, among all the nations of the earth, to attain full modernity; American intellectuals find themselves in the forefront of this movement. The tradition that I have described, in very brief outline, has been in large part their work; they are in turn one of the results of this tradition. The two missions of the modern intellectual are, first,

to investigate, create, and transmit knowledge, values, and experiences; and, second, to criticize society and its uses, institutions, and politics. Since the eighteenth century, this second
function, inherited from the medieval clerics, has assumed
greater and greater importance. We are all familiar with the
work of Americans in the fields of the sciences, literature, the
arts, and education; they have also been honest and courageous in their criticism of their society and of its defects. Freedom of criticism and of self-criticism has played a decisive role
in the history of the United States and in its present greatness.
Its intellectuals have been faithful to the tradition upon which
their country was founded and in which the scrutiny of conscience occupies a central place. This puritan tradition, however, by emphasizing and encouraging separation, is antihistorical and isolationist. When the United States abandons
its isolation and participates in the affairs of the world, it does
so in the manner of a believer in a land of infidels.

American writers and journalists have an insatiable curiosity
and are extremely well informed about what goes on in today's world, but instead of understanding, they pass judgment.
It must be said, in all truth, that they reserve their severest
judgments for their compatriots and those in public office. That
is admirable; yet at the same time it is not enough. In the days
of their country's intervention in Indochina, they denounced,
with good reason, the policy emanating from Washington; yet
their criticism, based almost exclusively on moral grounds,
generally neglected to examine the nature of the conflict. Critics were more interested in condemning Johnson than in understanding how and why there were American troops in
Indochina. Many said that this conflict "was no concern of
America's," as though the United States were not a world power
and the war in Indochina were a local episode.

Isolationism has been, alternately, an ideological weapon of
conservatives and of liberals: in the days of Franklin Roosevelt
it was used by the former, as today it is used by the latter.

Morality is no substitute for historical understanding. That is precisely why many liberals were so suprised at the outcome of the conflict: the installation of a military-bureaucratic dictatorship in Vietnam, the mass murders under Pol Pot, the occupation of Cambodia and Laos by Vietnamese troops, the punitive expedition by the Chinese, and, in recent days, the hostilities between Vietnam and Thailand. And today, confronted by the situation in Central America, liberals mouth the same simplistic nonsense. . . . Apart from the fact that it is not always sincere (often it is a mask), the moralizing attitude does not help us to understand the reality that lies outside ourselves. Nor does empiricism, or the cynicism that power brings. Morality, in the sphere of politics, must be accompanied by other virtues. Central to all of them is historical imagination. That was the faculty shared by Vico and Machiavelli, by Montesquieu and Tocqueville. This intellectual faculty has a counterpart in the realm of sensibility: sympathy for the other, for others.

The image presented by the United States is not reassuring. The country is disunited, torn apart by dissensions that do not have the least element of grandeur, eaten away by doubt, undermined by a suicidal hedonism, dazed by the ranting of demagogues. A society divided, not so much vertically as horizontally, by the clash of tremendous selfish interests: great corporations, labor unions, "the farm bloc," bankers, ethnic groups, the powerful communications industry. Hobbes's image has become an altogether concrete reality: everyone against everyone. The remedy is to regain unity of purpose, without which there is no possibility for action—but how? The malady of democracies is disunity, mother of demagogism. The other road, that of political health, leads by way of soul searching and self-criticism: a return to origins, to the foundations of the nation. In the case of the United States, this means to the vision of its founders—not to copy them, but to begin again. Not to do exactly as they did but, rather, like them, to make a new beginning. Such beginnings are at once purifications and mutations: with them something different always begins as well.

The United States was born side by side with modernity, and now, in order to survive, it must confront the disasters of modernity. Our era is a terrible one, but the peoples of the Western democracies, with Americans at their head, anesthetized by nearly half a century of prosperity, have stubbornly closed their eyes to the great blot that is spreading all over the planet. Beneath the mask of pseudo-modern ideologies, old and terrible realities are making their way back into our century, realities that the cult of progress and the mindless optimism of abundance believed were buried forever. We are living through a veritable revolution-and-return of times. More than a century ago, in the face of a situation less threatening than ours, Melville wrote lines that deserve to be read and pondered by Americans today:

When ocean-clouds over inland hills
Sweep storming in late autumn brown,
And horror the sodden valley fills,
And the spire falls crashing in the town,
I muse upon my country's ills—
The tempest bursting from the waste of Time
On the world's fairest hope linked with man's foulest crime.
Nature's dark side is heeded now. . . .

III

The Totalitarian Empire

The Cyclops and Its Flocks

From the moment of its birth in 1917, the true historical nature of the Soviet Union has been a matter of dispute. The first to question whether the new regime was in fact a "dictatorship of the proletariat," in the Marx and Engels's sense, were the Mensheviks and the European Marxists, the Germans and the Austrians in particular. The anarchists, for their part, immediately denounced the regime as being a capitalist state dictatorship. Neither Lenin nor Trotsky ever maintained, as Stalin was later to do, that the Soviet Union was a socialist country. According to Lenin, it was a transitional regime: the proletariat had taken power and was preparing the bases for socialism. Lenin, Trotsky, and the other Bolsheviks hoped that the revolution of the European working class, above all in Germany, would finally fulfill the prophecy of Marx and Engels: socialism would be born in the industrial countries of the West, those that were most advanced and had a working class that was heir to a tradition of democratic struggles. In 1920, however, in a speech criticizing Trotsky in the harshest and most vehement terms, Lenin said: "Comrade Trotsky speaks of a workers' state. That is an abstraction! In 1917 it was quite natural

that we should speak of a workers' state, [. . .] but today our
state is instead a state with a bureaucratic deformation. This is
the deplorable label that we must pin on it. . . . The proletariat, in the face of such a state, must defend itself. . . ." Words
uttered in 1920, words that today, in 1980, after the strikes
and the repressions in Hungary, Czechoslovakia, and Poland,
have lugubrious overtones.

Years later, Trotsky took up Lenin's criticism of him and made
it his own. In 1936, in the very midst of the battle against
Stalin and his theory of "socialism in one country" (an incon-
gruity from the point of view of true Marxism, though one
that has been parroted by thousands of intellectuals who call
themselves Marxists), he published *The Revolution Betrayed*. This
was the first serious attempt to decipher the true nature of that
new historical animal, the Soviet State. To Trotsky, it represented "an intermediate society, between capitalism and socialism," and one in which "the bureaucracy has turned into
an uncontrolled caste, alien to socialism." Trotsky thought that
the social struggles within it would resolve the ambiguity of
the degenerated workers' State in favor of one or the other of
its two conflicting tendencies: either the restoration of capitalism by the bureaucracy or the overthrow of the bureaucracy
by the proletariat and the inauguration of socialism. He was
thus unwilling to accept the idea that the domination of the
bureaucracy could continue *without* its relapsing into capitalism. Nonetheless, shortly thereafter, in his polemic with Max
Schachtman and James Burnham (1937–1940), he eventually
admitted, reluctantly, the possibility that the situation might
go on indefinitely, in which case the bureaucracy might form
a new class of oppressors and institute a new sort of exploita-
tion. He compared this eventuality, not without reason, to the
coming of the Dark Ages at the eclipse of antiquity.[1]

1. See Leon Trotsky, *In Defense of Marxism* (1942). The French edition (*Défense
du Marxisme*, 1976) is more complete.

In 1939 an Italian Marxist, Bruno Rizzi, published a book entitled *La Bureaucratisation du monde*, seldom cited but frequently plagiarized. In an embryonic form, more intuitive than scientific, Rizzi was the first to postulate, not as a remote possibility but as a visible reality in the Russia of Stalin and in the Germany of Hitler, the idea of a new regime that would succeed capitalism and bourgeois democracy: bureaucratic collectivism. A hypothesis that has met with good fortune: James Burnham adopted it, and it was later formulated and developed independently by Milovan Djilas (*The New Class*, 1957). The essays by Kostas Papaioannou published in *Le Contrat Social* and in other periodicals, though less empirical than Rizzi's and Djilas's, were written within the great tradition of modern historiography and are a notable contribution to the study of the genesis of the new class. In his book on postindustrial society, Daniel Bell devoted a number of penetrating and enlightening pages to the subject. The historical studies of François Fejto are also noteworthy. In the area of the philosophical inquiry into the nature of communism, the work of Leszek Kolakowski is of major importance; in historical and political sociology, Raymond Aron's analyses have cleared the path and lit the way for all of us. Other authors—Wittfogel, Naville, Bettelheim—have pondered the peculiarities of Soviet economic structures: do they represent State capitalism, bureaucratic monopoly, an Asiatic mode of production?

For Hannah Arendt and, more recently, for Claude Lefort, the real novelty is a political one: in all of history there has never been anything similar to the modern totalitarian system. In fact, the only comparable examples are remote societies such as the Egypt of the Pharaohs and, most important, the Chinese Empire. Etienne Balazs's studies (*La Bureaucratie céleste*, 1968) are doubly illuminating: on the one hand, the prolonged domination of the mandarins proves that a bureaucratic regime, contrary to what Trotsky thought, is not simply a transitory state and may last not just for decades but for centuries, mil-

lennia; on the other hand, the differences between the Chinese
Empire and the Soviet Union are enormous, and almost all of
them favor the former. Alain Besançon emphasizes the privileged function of ideology within the system—it is an illusory
reality, yet one more real than true, down-to-earth, humble
reality itself—and proposes that the system be labeled an
"ideocracy." Cornelius Castoriadis stresses the dual nature of
bureaucratic capitalism: it is a society of castes dominated by
an ideological bureaucracy, and it is a military society. Russia,
Castoriadis maintains, has passed imperceptibly from a regime
of domination by the Communist Party to another, in which
military realities and considerations are primordial, which he
therefore calls a "stratocracy" (Greek *stratos* = army).[2]

The list of interpretations and the assignment of descriptive
names could go on indefinitely, but the truth of the matter is
that there is agreement underlying this taxonomic quarrel. No
serious author today, in 1980, maintains that the Soviet Union
is a socialist country, nor that it is, as Lenin and Trotsky believed, a workers' state deformed by a bureaucratic excrescence. If we think of it in terms of institutions and political
realities, it is a totalitarian despotism; if we take a look at its
economic structures, it is a vast State monopoly with peculiar
forms of transmission of the use, the enjoyment, and the benefits of wealth and the products of labor (not the ownership of

---

2. The subject, as can be seen by this incomplete recapitulation, is immense
and continues to give rise to endless reflections. For instance, just as I was
about to send these pages off to the printer's, I received Edgar Morin's *De la
nature de l'URSS* (1983). Morin sees Russian totalitarianism as being a system
of superimposed dominations, each of them incorporating the next smaller
one in the manner of Chinese boxes: the State takes over civil society, the
party the State, the political committee the party, and the apparat (the Secre-
tariat) the committee. At the very top the domination is dual: the police keep
watch on the apparat, and the apparat controls the police. The apparat does
not coincide precisely with the bureaucracy: it is not a class *within* the State
but *above* it. On taking over the nation, the apparat also adopted nationalism
and Russian imperialism. Hence, the Soviet Union is on the one hand a totalitarianism, and on the other, without contradiction, an imperialism.

property, but what is equivalent to holding stock in a capitalist society—namely, being listed as belonging to the *nomenklatura* or being a card-holding member of the Russian Communist Party); if we take note of its social divisions, it is a hierarchical society with very little mobility, in which classes tend to become petrified as castes, dominated at the top by a new category at once ideological and military: an ideocracy and a stratocracy. This last description is particularly apt: the Soviet Union is a society fashioned in the image and likeness of the Communist Party. And the models for this party were the Church and the army; thus its members are clerics and soldiers, its ideal of community the cloister and the barracks. The cement binding together the religious and the military order is ideology.

Even though it appears to be a forbiddingly solid mass of ice and iron, the Soviet Union is faced with contradictions not less but more profound than the American Union. The first of these is basic, deeply imprinted in its very nature. Russia is a hierarchical society of castes and at the same time it is an industrial society. Because it is hierarchical, it is doomed to immobility; because it is industrial, it is doomed to change. Its social mobility is very nearly nonexistent, but the industrial transformations that it has undergone, above all in the area of heavy industry and military technology, are notable. In Russia it is things that change, not people; hence the immense cost of industrialization, in lives and in human labor. The inhumanity of industry, a feature that characterizes all modern societies, is accentuated in the Soviet Union because production is primarily oriented not toward satisfying the needs of the population but toward furthering the policy of the State. What is most real—human beings—is placed in the service of an ideological abstraction. This is a form of alienation that Marx did not foresee.

On the one hand, social and political fossilization; on the other, continual technical and industrial renewal. This contra-

diction, the source of injustice and inequality, brings about
tensions that the State stifles with methods common to all dictatorships: the strengthening of the system of repression, and
a policy of outward expansion. Empire and police: these two
words reveal that, despite the considerable differences that
separate them, there is a clear historical continuity between
the bureaucratic State and the czarist. Possessed by an ideology
no less expansionist than the old Pan-Slav messianism, the
Russian State has created a powerful war machine fed by a
gigantic military industry. Among all the inequalities of this
society, perhaps the most striking is the disproportion between
the living standards of the people—notably low, even compared with that of Czechs, Hungarians, and Poles—and the
enormous military power of the State. And so two contradictory definitions of the Soviet Union are equally true; that of
the poet Hans Magnus Enzensberger, who sees in real socialism the highest stage of underdevelopment, and that of Castoriadis, who defines it as stratocracy. With a remarkable
ignorance of the recent past, there has been renewed discussion in cafés and in university circles as to whether Russia is
or is not socialist. Engels anticipated both the question and the
answer long ago when he called the State capitalism of Bis-
marck "barracks socialism."

Although there has been since 1920 a flood of books and
reports on the true reality of Russia, many in the West and in
Latin America—intellectuals in particular, but also more than
a few liberal and conservative politicians, progressive-minded
bourgeois, Catholic clergy and laymen of the left—chose for
years on end to close their eyes. The Khrushchev Report blew
the lid off, making the reality impossible to ignore. Shortly
thereafter, the first texts of the dissidents surfaced. Since then
it has no longer been possible to affect ignorance. More fortunate than Pascal in his polemic against the Jesuits, Solzhenitsyn has succeeded in moving the world. His influence has even
converted Parisian literary-philosophical *cénacles*; in less than

five years we have been witness to the abandonment of the countless varieties of Marxist scholasticism previously predominant in European universities. Even the *"précieuses ridicules"* have ceased to quote Marx's *Grundrisse* and *Das Kapital*.

Nonetheless, however great the influence of the dissidents—Russians, Poles, Czechoslovakians, Rumanians, Hungarians, Cubans—outside their own countries, their possibilities for action within their homelands are extremely limited. The dissidents have proved that there is an abyss between true reality and ideological reality, and their descriptions have pinpointed it precisely; their diagnoses have been less accurate, however, and their remedies have not even been tried.

It is not easy to determine what course the evolution of Russian society will follow. But it is quite easy to predict that the contradiction that I have all too briefly described will become more and more pronounced in the immediate future and will become more acute once the generation of septuagenarians who are the leaders of the Soviet Union today passes away. The dissident intellectuals whose existence is well known in the West are simply a political and religious manifestation of the basic contradiction. Anyone who has had personal contact with Russian intellectual life in the universities and scientific centers of that country—or those of the satellite countries—discovers immediately that the official ideology, Marxism-Leninism, has become a catechism that everyone recites but no one believes. The erosion of State orthodoxy is one aspect of the divorce between reality and ideology; other evidences of this contradiction are the worrisome reappearance of Pan-Slavism, the nostalgia for czarist autocracy, anti-Semitism, and Great Russian nationalism. The past of Russia is alive and is returning.

At a deeper level than these ideological tendencies, other aspirations are making themselves felt, although as yet they have not found any channels of expression—demands that are broader and more concrete than those of the intellectuals. All visitors to the Soviet Union have observed the eagerness of the urban population to adopt Western life styles, especially American ones. It is not an exaggeration to speak of the "Americanization" of the young people in the largest cities. The fascination with Western society is not limited to imitating its most deplorable manifestations, such as "consumerism." It is scarcely necessary to call to mind the fact that the Russian worker lacks such basic rights of labor as the right to strike, to organize, to join a union, to attend union meetings. Is it possible to create a powerful industrial State with a passive and demoralized proletariat, whose only ways to fight for its rights are alcoholism, loafing on the job, sabotage? How will the fossilized Russian regime confront the double pressure of those demanding more freedom (the intellectuals) and those demanding more and better consumer goods (the general population)?

The revolt of the Polish workers is destined to have an immense influence, in Russia as well as in the satellite countries. It does not matter that the Polish army put down the rebellion in Poland: what counts is that from the Kronstadt rebellion, crushed by Lenin and Trotsky in 1921, to the strikes in Poland in 1981, there has been an unbroken series of popular uprisings and riots against the communist bureaucracies. We do not receive news of disturbances in the Soviet Union—though the accounts by Solzhenitsyn and other dissidents have dispelled our ignorance somewhat—but the events in Hungary, Czechoslovakia, and Poland are on everyone's mind, as is the flight of the hundred thousand Cubans via Puerto Mariel.

In addition to this internal contradiction, the Russian bureaucratic system is faced with another, which, although it manifests itself within Soviet borders, must nonetheless be called external. Like the United States, the Soviet Union comprises groups of quite different origins. The population of the United States is made up of immigrants (except for the indigenous Indians and a very small sector of the population that is of Mexican origin), who were subjected to that system of integration and assimilation known as the "melting pot." The experiment produced notable results: the United States is a country with traits all its own and a marked originality. Blacks, Chicanos, and others were admittedly left out of the "melting pot"; moreover, it did not entirely dissolve the national characteristics of the different groups dumped into it. Still, it is evident that the process of unification and integration is by now very far advanced and irreversible. All citizens of the United States, even those who are most discriminated against, feel that they belong to the same country, speak the same language, and, by an overwhelming majority, profess the same Christian religion.

Czarist expansion occupied many territories militarily and subjected their populations to Russian imperial rule. Although the situation has changed, these nations have not disappeared: the Soviet Union is an entity made up of different peoples, each with its own language, culture, and predominating religion. Within this whole, Russia properly speaking (the Russian Soviet Federated Socialist Republic) dominates the others. Hence two features characterize the Soviet Union from the point of view of nationalities: heterogeneity and domination. The Soviet Union is an empire in the classic sense of the word: a group of scattered nations not related to one another—each one with its own language, culture, and tradition—subject to a central power.

The national tensions within the Russian Empire, as is well known, are frequent and permanent. The nationalism of the Ukrainians is still alive, despite persecutions; the same can be said of the Balts, the Tatars, and the other nations. According to the French writer Hélène Carrère d'Encausse (*L'Empire éclaté*, 1979), national contradictions are destined to play an even more crucial role in the Soviet Union than social and ideological ones. She may well be right: the history of the twentieth century has been the history not only of class struggle but also of warring nationalisms. The case of the Soviet nations that profess the Moslem faith has a special significance. These are peoples who have kept their national and cultural identity; their demographic growth has been extraordinary, and in a few years they will constitute two-fifths of the Soviet population. As we know, the renaissance of Islam, a militant religion, has left its profound mark on our era; it is impossible to believe that it will stop at the gates of the Soviet Union. The Russian bureaucratic State has not managed to resolve either the national or the religious question, which for Islamic tradition are one and the same. In a not too distant future, the government in Moscow will be obliged to confront, within the frontiers of the Soviet Union, the triple danger of Islam: religious, national, and cultural.

Within the area ruled by these two great contradictions—the social and economic one, and the ethnic and religious one—others, of a linguistic, cultural, and political order, proliferate. They are an accumulation at once complex and explosive. The Russian State has kept itself from being blown apart by the two means habitually resorted to in all dictatorships: repression and the shifting of the locus of conflicts to the world outside. The terror of the Stalin era was unique in history and can be compared only to that of Hitler, his contemporary and rival. The great exterminators of the past—Genghis Khan, Attila, Tamerlane, the Assyrian monarchs who devastated Asia Minor in the Terror Assyriacus—are modest figures alongside these two scourges of the twentieth century.

It is impossible to ignore the influence that terror has had in the taming of the Russian public spirit. After Stalin there was a letup—Khrushchev's policy of reforms—but it went only halfway and lasted for just a short time; Brezhnev froze it solid, and the Russian regime continues to be a rule through despotism and police repression, albeit without the excesses of Stalin. It is not easy, however, to liberalize the regime without endangering the dominant caste and its privileges. In Russia there is no such thing as that free political space—the arena where classes and groups confront one another, advance, retreat, and reach compromises—that for more than a century has enabled workers to win the battles that they have won. The *nomenklatura*—as the privileged class is called in Russia—counters the increasing social pressure with a rigidity that is also increasing. Thus Russian society lives beneath a twofold threat: petrification or explosion.

For more than ten years now, the Soviet government has been pursuing a frank policy of expansion. This movement, on the one hand, is the consequence of the errors and vacillations of U.S. foreign policy; on the other hand, it is the safety valve for internal conflicts and tensions. In the coming years, in the face of the uncontainable nature of its social and national contradictions, the Russian government is bound to seek expansion toward the outside as a way out. Until now, political expansion has been accompanied by military occupation—or else, in such cases as Poland, Czechoslovakia, Cuba, and Vietnam—it has made the survival of the government dependent upon Russian military aid. The Soviet Union has returned to the old conception of imperialism, which identified domination with direct power over territories, governments, and populations. The paradox is that the Soviet Union does not need either the territories or the natural resources of other countries (it needs their technology, but this it can obtain easily through credits and trade with Europe, Japan, and the United States). Thus the principal aim of Soviet expansion is to export its internal contradictions to the world outside its borders. And to export the imperial ideology—or, as the Chinese put it, the "chauvinism" of great power—inherited from czarism and combined with Marxist-Leninist messianism.

To these circumstances we must add another, pointed out by Cornelius Castoriadis in a luminous essay (*Devant la guerre*, 1981). As I indicated earlier, according to Castoriadis the Soviet Union has turned into a stratocracy; the communist bureaucracy, using the weapon of ideology, imposed a reign of terror on civil society; but today, as we all know, ideology has evaporated (leaving cynicism, venality, and hypocrisy as residua) in the Russian social conscience and in practice as well. This ideological vacuum has been occupied, according to Castoriadis, by considerations of a military nature, and, consequently, the Soviet army tends more and more to take the place of the party. The military society is a society *within* Russian society. In the broad sense of the word "society"—that is, in the sense of a technical, scientific, economic, and industrial complex—"the army is the only truly modern sector of Russian society and the only one that functions effectively." The Russian military State, like all military States, knows how to—is able to—do only one thing well: make war. The difference as compared with the past is that the old military States did not have nuclear arms in their arsenals.

Allies, Satellites, and Rivals

The relationship of the United States with its friends and its clients has become critical, in consequence both of its most recent failures and of the nature of its domination. The term "imperialism" is used to characterize the latter, but the term "hegemony" fits it better. Dictionaries define "hegemony" as the *supremacy* of one State over others, "supremacy" being used in this case in the sense of "predominant influence." Empire, on the other hand, implies sovereignty not only over subject peoples but over their territories as well. The domination of Latin America by the United States has been of the nature of a hegemony; it has almost never been exercised directly, as in the case of empires, but, rather, through its influence on governments. That influence, as is well known, is different in each instance and leaves a more or less ample margin for negotiation. Although the United States has not hesitated to intervene militarily on many occasions, such acts of intervention have always been seen as violations of law.

In the beginning of its history, the United States admittedly resorted to typically imperialist-military expansion and appropriated territories that were Mexican or were still under Spanish rule. But since early in this century, U.S. hegemony has had objectives that were primarily economic, and only secondarily political and military. In recent years the situation has changed: again and again the United States has been obliged to accept governments in Latin America that were not to its liking. At one extreme, Cuba; at the other, Guatemala. In between, a spectrum ranging from Nicaragua to Chile.

On the other continents, the situation is no different. Western Europe and Japan have shown that they are partners, not servants, of Washington. The international policy of France has traditionally been nationalistic, and its independence in the face of the United States might even be termed fastidious. Since Brandt, the Federal Republic of Germany has tried to go its own way. It is an open question whether the policy of the German and French governments has been prudent or whether it has been inspired, at least in part, by the anti-Americanism that prospers among French nationalists as well as among the socialists and social democrats of the two countries. Certain people on the right have invoked the precedent of General De Gaulle, who years ago embarked upon a similar policy. His objectives were different, however: insofar as possible, it was his intention to re-establish the balance of powers. Those were the years of American superiority, whereas today the relationship of forces has changed. But what is beyond question is the fact that the foreign policies of Paris and of Bonn are founded on a frank evaluation, right or wrong, of the national interests of France and of the Federal Republic of Germany.

In the Middle East and in Asia conditions are analogous. The relationship between the United States and Israel is *sui generis* and cannot be reduced to the oversimplified explanation that it is merely one of dependence. The Israelis have as great a need of American arms as the presidents of the United

States have of the votes and the influence of American Jews. The situation is similar with regard to Egypt and Saudi Arabia: in each case the needs of both parties are mutual and recipro- cal. With Pakistan on one border and China on another, India cultivates an independent policy, frequently closer to Moscow than to Washington. Pakistan is not entirely dependent upon the United States, and for a long time now has had a special relationship with China. As for the nucleus of the system—the United States, Western Europe, Japan, Australia, and Can- ada—this alliance is based on self-interest and a consensus as to the value of certain institutions and principles, such as rep- resentative democracy, respect for minorities, and human rights. This consensus, however, is not an orthodoxy. From all the foregoing, we may conclude that the problem for the interna- tional policy of the United States is the same as for its domestic policy: how to find, within the plurality and diversity of wills and interests, a unity of purpose and a unity of action.

The relationship of the Soviet Union to the countries that belong within its orbit is altogether different. The relationship has political, military, and ideological aspects, all of them re- flections of a single reality. These countries are united by a single doctrine, the canonical version of which is that of Mos- cow, the central power. It is true that the Russian State has become a little more tolerant than during the Stalin era, that it permits Rumania to assume defiant stances and set aside its excommunication of Tito; nonetheless, the margins for inter- pretation of the doctrine continue to be extremely narrow, and any political difference immediately becomes heresy. As in the theocracies of antiquity, the communist system has succeeded in fusing power and idea. Thus all criticism of the idea be- comes conspiracy against power; disagreement with power, sacrilege. Communism is doomed to engender schisms, to cause them to multiply, and to repress them.

Orthodoxies with universalist and exclusivist pretensions tend successively to sectarianism and to its persecution. The precedent of Christianity is instructive. The Church-State of Constantine and his successors kept the doctrine from breaking up into hundreds of sects, but the cost was enormous: the theological State was also the inquisitional State. With even greater fury than the bishops and the monks, the Soviet State has persecuted all deviations. Its relations with the satellite governments reproduce this theocratic conception of politics. Each of the satellite governments in turn postulates a version of the doctrine that is equally canonical and universal—within its own borders. The doctrine, like the image in the shards of a broken mirror, is thus multiplied, and each fragment makes itself out to be the original, the sole, the authentic version. Universality is always in danger of identifying itself with this or that national version. Moscow has succeeded, up to a certain point, in preventing the proliferation of heretical versions through the combined use of flattery, intimidation, and, when necessary, force. If a communist State depends substantially on Moscow for military and economic aid, like Cuba and Vietnam, the problem of orthodoxy does not even present itself. The same is true when governments, in order to keep themselves in power, require Soviet tanks: Hungary, Czechoslovakia, Poland.

The case of Afghanistan reveals the tendency of the Russian State to make ideology the very substance of politics. (A curious perversion of idealism: only ideology is real.) The Russian adventure in Afghanistan is the mirror image of the policy followed by the British Empire. Throughout the past century, the English tried to dominate the Afghans; although they never succeeded in dominating them completely, they at least prevented the country from falling into the hands of czarist Russia. The only government that had the right to keep a diplomatic mission in Kabul was Great Britain. But it never entered the minds of the English to convert the Afghans either to the Anglican religion or to constitutional monarchy. In 1919 Afghanistan regained its independence and opened its doors to the world. Then, after World War II, the liquidation of the British

Empire precipitated events. The United States took the place
of the English, but they were not able to contain the Russians
for long. The historical circumstances had changed radically,
and besides, Washington defended its position halfheartedly;
it never regarded Afghanistan as a key strategic point. A grave
error: ever since Alexander the Great this country has been the
gateway to the Indian subcontinent.

The Soviet government had better luck than the czarist, and
managed to infiltrate more and more circles within the country, above all groups of young army officers. The domestic policy of Afghanistan favored the Russians. Mohammed Zahir Shah,
the king of Afghanistan, occupied the throne in 1933, after the
assassination of his father. Because he was so young, his close
relatives acted as regents, above all his brother-in-law Daoud
Khan, who soon took over as prime minister and strongman
of the regime. The king, eager to free himself from the tutelage
of Daoud, headed a peaceful revolt of notables. Afghanistan
became a constitutional monarchy, and by law the kinsmen of
the monarch were excluded from government. The task of
running the country was turned over to representatives of the
enlightened faction, which was determined to transform the
country into a modern society. In matters of international policy, Zahir Shah's regime, under the leadership of Prime Minister Hashein Maiwandwal and later under the no less intelligent
Ahmed Etemadi, both liberals, was strictly neutralist. It was
the era in which the nonaligned movement had not yet turned
into an agency of Soviet propaganda.

Daoud was a man of the past. Astute, cruel, and single-
minded, he sought friendship with the Soviet Union. He allied
himself with pro-Soviet army officers (almost all of whom had
studied in Russian military academies), and, in a bloodless coup
d'état, Zahir Shah was dethroned. The country was declared a
republic and Daoud was named president. Daoud's coup con-
solidated the predominance of Russia in Afghanistan and put
an end to what remained of American and Western influence

in the country: Afghanistan was turned into a Finland of the East.

But this political and strategic victory was not enough to satisfy the Soviet government. Once again ideology appeared on the scene: in order to make Russian domination total, it was necessary to transform Afghanistan—a deeply religious country, divided into kinship groups, tribes, and rival fiefs—into a People's Republic. Another coup d'état put an end to Daoud's power—and to his life. What happened thereafter is a familiar story: the struggle between ideological factions, terror and its thousands of victims, two more coups d'état, both bloody, generalized rebellion against the communist dictatorship, and armed intervention.

There are those who have compared the role played by Russia in Afghanistan to that played by the United States in Vietnam. A misleading comparison: the differences are enormous. The Americans lent their support in a war that was profoundly unpopular all over the world, including in their own country; they fought in a territory thousands of miles away from their own, against an enemy supplied with arms and equipment by two great powers, Russia and China; the North Vietnamese cause, moreover, was supported by a powerful current in world opinion. The Russians, on the other hand, are operating in a country bordering on theirs, against an isolated enemy that lacks organization and is poorly armed; the Russian government does not have an independent public opinion at home to which it must account for its actions, nor is it confronted with a wave of international censure: where are the intellectuals, the clergymen, the students to demonstrate in support of the unfortunate Afghans as they did for the Vietnamese?

The Russian intervention in Afghanistan is yet another example of a well-known fact: the Soviet Union follows, in its broad outlines, the foreign policy of the czarist regime. As is always the case, this policy is dictated by geography, the mother of history; it is also a continuation of an expansionist imperial tradition. In the past, the ideology that supported this expansionism was Pan-Slavism; today it is Marxism-Leninism. The change in ideology parallels another, which is material and historical: Russia has gone from being a European power to being a world power. This destiny has awaited her since the beginning, as was foreseen by a number of clear-minded thinkers of the nineteenth century, among them a Spaniard: Donoso-Cortés. Eighty years ago, Henry Adams predicted that soon two opposing forces would confront each other: Russian inertia and American intensity. He was not mistaken, though would he be able to speak today of inertia in the case of Russia? The terms that better describe her policy are "tenacity" and "patience." In another passage in his autobiography, Adams saw the future course of events with absolute clarity: "Russia must fatally roll—must, by her irresistible inertia, crush whatever stands in her way." Yes, a steam roller . . . The change in ideology has modified neither the deep character of this great nation nor its government's methods of domination. From the nineteenth century on, imperialism was no longer ideological: it was a political, military, and above all an economic expansion. Neither the English nor the French nor the Dutch seriously tried to convert their colonial subjects, as the Moslems and the Catholic Spaniards and Portuguese before them had done. Moscow is returning to the old fusion between power and idea, as in premodern empires. Russian bureaucratic despotism is an imperialist ideocracy.

Milan Kundera is shocked by the wide use of the expression "Eastern Europe" to designate those European countries governed by Russian satellite regimes; these countries, the Czechoslovakian novelist says, are a part of Western Europe and are undergoing a situation that at bottom is not very different from what they experienced during the war, under puppet governments forced on them by the Nazis. He is right. It should be added that the Russian domination over these peoples has retarded their historical development. Czechoslovakia, for instance, was already a wholly modern country, by virtue not only of its notable scientific, technological, and economic development, but also of the beneficence of its democratic institutions and the richness and vitality of its culture. In Czechoslovakia, Rumania, Hungary, and Poland (the case of Bulgaria is different, and the East German communist regime is simply the result of the Russian occupation), liberation from the Nazis was achieved with the participation of Soviet troops. In Rumania and in Hungary it was the Russian military that led the way to the establishment of a communist regime; in Czechoslovakia and Poland, where there were popular anti-Nazi and democratic movements, the process was more complex, but the influence of Moscow was the decisive factor in orienting these nations toward communism. Since that time, the shadow of the Soviet Union has fallen over the entire territory of all these countries; its leaders are answerable first and foremost to Moscow, the central authority, and only secondarily to their peoples.

The governments of East Germany, Rumania, Czechoslovakia, Poland, Hungary, and Bulgaria reproduce the totalitarian model of the parent State, although there are, naturally, differences between them: Rumania's is less dependent on Moscow than Germany's and Czechoslovakia's, Hungary's is more liberal than the others, and so on. But in all of them the features that define communist ideocracies are present: fusion of the State and the party, or, better put, the takeover of the State by the party; the political and economic monopoly of the bureaucratic oligarchy; the preponderant function of ideology; the transformation of the bureaucracy into a stratocracy, in the meaning that Castoriadis has given this term. The characteristic note, common to all these regimes, is their dependence on the outside: the bureaucratic-military caste maintains itself in power thanks to Russian troops.

The struggle that European peoples have been waging for years against Russian domination and governments imposed

by Moscow is of a special nature. In the first place, these countries have a very rich and very old national past; all of them
except Bulgaria have managed to enter the Modern Age, economically and culturally. A number of them, Czechoslovakia
in particular, governed themselves through advanced democratic institutions at a time when other European nations, such
as Italy and Germany, were ruled by fascist regimes. It is equally
significant that the struggle against the communist oligarchy
was initiated by the working class. If the recrudescence of class
struggle in so-called socialist regimes is scandalous, what are
we to say of the repression that has been unleashed in all of
them by the communist oligarchy, with the backing of the
Soviet Union? It is also moving to think that Hungarian,
Czechoslovakian, and Polish workers are fighting for rights that
are recognized in the rest of the world, even in Latin American
countries under reactionary military dictatorships. And there is
something else: the struggles for freedom of association and
for the right to strike are merely one aspect of a vaster movement that includes a country's entire population. What is taking place is a struggle for basic freedoms. The battle for
democracy, in turn, is inseparable from the movement for national independence. The revolts against communist oligarchies are aimed not at restoring capitalism but at re-establishing
democracy and regaining independence. What has been sought
in the uprisings in Hungary, Czechoslovakia, and Poland is national resurrection.

Not a few European and Latin American intellectuals attempt to equate the policy of the United States with that of the
Soviet Union, as though they were twin monsters. Hypocrisy,
naïveté, or cynicism? It seems to me that what is monstrous is
the comparison itself. The errors, the failures, and the sins of
the United States are enormous, and I am not trying to absolve
that nation. Nor do I excuse the other Western capitalist democracies and Japan. The policy of all these governments toward Russia has been incoherent and almost invariably one of

weakness, though interrupted by aggressive stances that have been sheer rhetoric; their blindness to the social and economic problems of the nations of Africa, Asia, and Latin America has been as great as their selfishness; they have frequently been the accomplices of horrible military dictatorships, and at the same time they have been indifferent to genuine popular movements (the latest example has been their attitude toward the revolutionary leader and Nicaraguan democrat Edén Pastora, known as Commandant Zero).

All this having been said, however, it must be added that the capitalist democracies have preserved fundamental freedoms within their own borders. On the other hand, ideological war abroad and totalitarian despotism at home are the two *constituent* features of the Soviet regime and of its vassal countries. There is no trace of either one of these two evils in the democratic countries of the West. And, despite endless adversities, seeds of freedom also exist in other countries of Asia and Latin America, such as India, Venezuela, Costa Rica, Peru, and Colombia. In Lebanon there used to be a democracy: will it be restored one day?

The governments of the West answered the Russian interventions in Hungary and Czechoslovakia with rhetorical protests and vague threats of sanctions. For the Soviet Union the dangers of its policy of force do not lie in the timid and self-interested reactions of the governments of the West, but in the reaction of the peoples subjected to their domination. Moscow strengthens the ties between the communist elites who exercise power but limits their ability to maneuver and shift position. What threatens to destroy American hegemony is disintegration. The greatest shortcoming of Russian hegemony is rigidity: hence the frequency of the explosions within its satellites.

On a few occasions the Soviet State has proved powerless to suppress heretical versions of doctrine. The reason: these versions had turned into the credo of independent States capable of standing up to Moscow. The first case was Yugoslavia, and Albania's situation is similar. The most serious and most decisive schism has been that of Peking. The conflict with China very soon degenerated into border skirmishes; it subsequently spread to Indochina. At first the two powers fought through their allies, Vietnam and Cambodia; after the Vietnamese aggression and the overthrow of Pol Pot, the Chinese launched a punitive expedition against Vietnam in order to teach it a "lesson." Marxism claimed to be a doctrine that would abolish the wage-slave class and bring about universal peace; the reality of contemporary communism offers us an image that is the diametrical opposite: servitude of the working class and quarrels between the "socialist" States.

Whatever the evolution of China may turn out to be, there is little likelihood that its differences with the Soviet Union will disappear in the foreseeable future. On the contrary, they will very probably grow more acute. Although the contenders maintain the contrary, the Sino-Russian quarrel is not ideological: it is a struggle between two powers, not between two philosophies. An extraordinary and a revealing evaporation of ideology—the ghosts of Machiavelli and Clausewitz must be smiling ironically. China feels that it is threatened by Russia; that is the heart of the question. Its fears are justified: not only does it share a tremendously long border with the Soviet Union, but among its other neighbors is an enemy country, Vietnam, and another with which it has traditionally had stormy relations, India.

The rivalry between China and Russia began in the seventeenth century. The Russians settled on the shores of the Amur River in 1650 and in the following year erected a fort at Albazin. In China, following a series of upheavals, the Ming dynasty was overthrown and the Manchu invaders enthroned a new dynasty (the Ch'ing) in its place. Occupied with pacifying the country and consolidating their rule, the Manchus were not able at first to risk a confrontation with the Russians. A quarter of a century later, things had changed.

In 1662 K'ang-hsi ascended the throne. He was a contemporary of three famous sovereigns: Aurangzeb, the great Mogul of India; Peter the Great of Russia; and Louis XIV of France. According to the historians, K'ang-hsi was not only the most powerful of the three but also the best governor, the most just and humane. He was a good soldier and an enlightened monarch, as well as a distinguished poet and a true scholar in the Confucian tradition. This latter trait did not keep him from taking a lively interest in the new European sciences. He became acquainted with them thanks to the learned Jesuit missionaries who lived in Peking in those years, under his patronage. In a book at once scholarly and scintillating, Etiemble recounts how the Jesuits intervened in the conflict between the Chinese and the "Muscovites," as chroniclers of the era called the Russians.[3] The first step K'ang-hsi took was to send troops to the Amur, take the fortress, and raze it. But the Muscovites returned with reinforcements, drove the Chinese out, and rebuilt the fort. Then K'ang-hsi sent for Father Verbiest, the Jesuit missionary who had made the instruments of the imperial astronomical observatory. The sovereign asked him to cast three hundred cannons; at first the priest refused, but later was obliged to yield in the face of the imperial request and over a year's time turned out the artillery pieces demanded. With the aid of these cannons the Russians were forced out, and a few months later, in 1689, the Treaty of Nerchinsk was signed between the emperor of China and the emperor of Moscow. It was negotiated by two other Jesuits—Father Gerbillon, a Frenchman, and Father Pereira, a Portuguese. Drafted in Latin, this document (Etiemble tells us) constitutes in large part the basis of the Chinese claims against the Soviet Union with regard to border questions. Chairman Mao was of the opinion that the Treaty of Nerchinsk had been "the last one between China and Russia signed on equal terms."

The United States and Russia are the centers of two systems

3. Etiemble, *Les Jésuites en Chine* (1966).

of alliances. A comparison between them reveals, once again, a symmetrical opposition. The system that allies the United States with the countries of the West and with Japan, Canada, and Australia is fluid and in constant movement. This is due, first, to the autonomy and relative freedom of action of each one of the members of the alliance; and, second, to the fact that all the States that compose it are democracies in which the ruling principle is the rotation of power among men, parties, and ideas: each change of government, in each country, results in a change, major or minor, in its foreign policy.

The Russian system is not fluid but fixed, and is founded not on the autonomy of the States but on their subjection. First of all, each of the allied governments depends on the central one, and since it is in power thanks to Russian military backing, it is not free to deviate from the line imposed by Moscow (although there are, naturally, degrees of subjection: Rumania enjoys greater freedom of movement than Bulgaria, and Hungary than Czechoslovakia); second, neither Russia nor the other members of the bloc follow the principle of the periodic rotation of men and parties, so that changes are far less frequent. Fluidity and fixity are a consequence of the nature of the political relation that unites the members of the two systems with their respective centers. The rule that the United States exercises can be defined, in the strict sense of the word, as a hegemony; that of the Russians, in the classic sense, as imperial dominion. The United States has allies; the Soviet Union, satellites.

The situation changes once we look at the periphery of the two systems. The United States experiences great difficulties in its dealings with the nations of Asia and Africa. Certainly there are countries on those continents that depend on the friendship and aid of the United States, but their relationship is not one of subjection. It cannot be said that the governments of Egypt, Morocco, Saudi Arabia, Indonesia, Thailand, Turkey—not to mention Israel and New Zealand—are satellites of Washington.

The difficulties of the United States stem in large part from the fact that it is looked upon as the heir of the European colonial powers. The most notable—and the most justified—case of this identification of Americans as the new colonial power was the Vietnam War. The situation in Latin America is even less favorable for the United States, both because of its interventions and intrusions in the affairs of our countries and because of its support for reactionary military dictatorships. Its relations with these latter have not always been harmonious, however, and many of the generals it has supported have turned out to be brazen, overbearing, and unruly. Finally, although it is a presence throughout the world by virtue of its wealth and its military strength, the United States does not aim at a universal orthodoxy, nor does it rely on having a political party in each country that regards it as the perfect incarnation of this ideology. Its entire policy has consisted in halting the advances of Russian communism, and this has turned it into a defender of the *status quo*. An unjust state of affairs. Neither the United States nor Western Europe has been capable of formulating a policy that is viable in the countries of the periphery. Both have lacked not only political imagination but sensitivity and generosity as well.

The Soviet Union is confronted with an entirely different situation. It is not looked upon as the heir of European imperialism in Asia and Africa; nor has it intervened in Latin America until recently, and even then (in Nicaragua and El Salvador) not directly but through the intermediary of Cuba. (The Russians and the Cubans deny any such intervention, which is understandable. What is less understandable is that a number of governments, among them those of France and Mexico, along with many intellectuals and liberal journalists—American, French, German, Swedish, Italian—should swallow this lie and spread it.) Another advantage which Russia enjoys is that certain sectors of public opinion in these countries, above all middle-class intellectuals, do not regard it as an expanding empire but as an ally against Yankee imperialism. The Soviet government, it is true, does everything possible to dispel illusions such as these, yet its acts in Hungary, Czechoslovakia, Poland, and Afghanistan have not succeeded in shaking the faith of these stubborn believers. To sum up in a few words what is the real heart of the matter: Moscow is the ideological and political capital of a belief that combines religious messianism and military organization. In each country the faithful, meeting together in parties that amount to militant churches, practice the same policy.

The democratic nations profess a superstitious veneration for change, which to them is a synonym of progress. Thus each new government proposes to carry out an international policy different from that of its immediate predecessor. To this periodic instability there must be added the recurrent appearance of a chimera: the possibility of arriving at a definite understanding with the Soviet Union. Again and again Westerners have believed, not in the possible—a *modus vivendi* that will prevent war—but in the impossible: a definitive division of spheres of power and of influence that will ensure, if not justice, at least universal peace. And again and again Russia has put an end to these accords with an act of force. The kind of changes that take place in the democratic countries are outside the experience of the Russians, who are not the victims of such illusions: their policy has been the same since 1920, and the modifications that it has undergone have not been fundamental or represented crucial shifts of direction, but instead are merely transitory tactical maneuvers. As Castoriadis observes with penetrating insight, the Russians do not want war, but neither do they desire peace—what they seek is victory. Russian policy is consistent, persevering, astute, and unyielding; moreover, it combines two elements that play a role in the creation of great empires: a national will and a universal idea. This way of going about things, based on firmness and shrewdness, patience and obstinacy, stands in sharp contrast to the oscillations and incoherencies of American policy and to the weary indifference of the great European nations.

If, instead of examining the *temper* of the two contending sides, we take a close look at the nature of their institutions and the internal conflicts that plague them, things come into clearer focus. In the United States and throughout the West, institutions were conceived in order to confront changes, channel them, and assimilate them; in Russia and its satellites, in order to prevent them. The distance between institutions and reality is very great in the West. In Russia this distance becomes contradiction: there is no relation between the principles that inspire the Russian system and social reality. The contradiction becomes more acute still as we go from the center to the satellites. The solidity of the Soviet Union is misleading: the true name of this solidity is "immobility." Russia cannot move; or, better put, if it moves it crushes its neighbor—or it cracks apart and collapses.

IV

Revolt and Resurrection

Quarrels on the Periphery

"Third World" is an expression that there is good reason to abolish. The label is not only inexact: it is a semantic trap. The Third World is many worlds, all of them different. The best example is the Nonaligned Movement, a heterogeneous group of nations united by a negation. The principle that inspired its founders—Nehru, Tito, and others—was a fitting one; its function, within the assembly of nations and in the face of the abuses of the Great Powers, should have been critical and moral. It has been neither, because the Nonaligned Movement has allowed itself to be carried away by ideological passions, and because practically none of the governments involved is in a position to offer lessons in political morality to other nations.

In reality, the only thing that has united these governments is the enmity that many of its members feel toward the West. An understandable sentiment: almost all these countries have been victims of European imperialism, either because they were colonies or because they suffered from European intrusions and exactions. It is also natural that they should not look with favor on the ally and head of the powers that once dominated them: the United States. The Soviet Union has astutely turned these feelings to its advantage and has tried, with increasing success, to maneuver the Movement into converting itself into a forum for attacks on the United States and Western Europe. The gravest error was to have admitted openly aligned countries, such as Cuba. If the Movement does not regain its former independence, it will not only lose its reason for being but, like the World Peace Council of the Stalin era, will end up a mere propaganda agency.

The end of European colonial empires and the transformation of their former colonies into new States can be looked upon as a great victory for human freedom. Unfortunately, many of the nations that have attained their independence have fallen beneath the domination of tyrants and despots who have become celebrated for their eccentricities and cruelties rather than for their genius as statesmen. The twilight of colonialism has not been the dawn of democracy. Nor has it been the beginning of prosperity; where democracy still survives, as in India, misery persists as well. Poverty is one of the scourges of the economically underdeveloped countries. Many are of the opinion that this is the principal obstacle preventing these peoples' attainment of democracy, but this is a half truth: India itself shows that underdevelopment and democracy are not altogether incompatible.

Nonetheless, it is clear that democratic institutions survive with difficulty wherever the process of modernization has not been completed, poverty reigns, and the groups constituting the base of modern democracies—the middle class and the industrial proletariat—are embryonic. Democracy is the consequence not of economic development, however, but of political education. Democratic traditions—the great English contribution to the modern world—have been better and more profoundly assimilated by India than by Germany, Italy, or Spain, not to mention Latin America.

The perversions from which Marxism has suffered in recent years oblige me to remind the reader that Marx and Engels always regarded socialism as a consequence of development and not as a method of reaching it. One of the lacunae of Marxism—I am referring to true Marxism, not to the wild elaborations upon it that circulate in our countries—is the poverty and the sketchiness of its concepts regarding economic underdevelopment. The great authors have said very little on the subject. All of them, beginning with Marx and Engels, had their eyes fixed on the most advanced capitalist countries. In any event, one thing is certain: the founders and their disciples—not excluding Lenin, Trotsky, and Rosa Luxemburg, not to mention Kautsky and the Mensheviks—always thought that socialism comes not *before* but *after* development. However, many intellectuals in the underdeveloped countries have believed, and still believe, that socialism is the quickest and most efficient way—and perhaps the only way—to emerge from underdevelopment. This fateful belief has been responsible for the appearance of historical hybrids that would have filled Marx with consternation and whose very name is a self-contradictory absurdity: "underdeveloped socialisms."

The idea of using socialism as an agent of economic development was inspired by the example of Russia. It is true that this country has become an industrial power, although the methods that Stalin used to achieve it were the negation of socialism. But there is nothing mysterious about this change. Tomorrow China, too, may be a great industrial power. Brazil will be one eventually, and in all likelihood so will Australia. The economic transformation of these countries does not have a great deal to do with socialism: they can all count on having the physical and human resources that a nation needs in order to become a world power. The amazing thing is that other countries, with fewer resources and fewer inhabitants, have reached, in less time and with less blood and tears, an impressive level of development: Japan, Israel, Taiwan, Singapore, etc. The same can be said of the differences in development among the European countries beneath Russian domination—East Germany, Czechoslovakia, Hungary, Poland, Rumania—

which are attributable not to different varieties of socialism but to different economic, technical, cultural, and natural circumstances.

Cuba is the proof that socialism is not a universal cure-all for economic underdevelopment. I cite Cuba because a number of factors make it an exemplary case. First, it has a more or less homogeneous population, despite a racial plurality, distributed over a limited, contiguous territory, with a standard of living and likewise of public education that even before Castro took power was one of the highest in Latin America (only slightly below that of Uruguay and Argentina). Second, unlike Vietnam and Cambodia, Cuba did not experience a devastating war. The American embargo has worked hardships on the Cuban economy, but trade with the rest of the world, and Soviet aid above all, have partly compensated for this setback. Finally, since the Castro regime has been in power for a quarter of a century now, it is possible at this point to judge where, in the long run, it has succeeded and where it has failed. It was said many times, and for many years—especially by Marxists—that the basic reason for Cuban underdevelopment and for its dependence on the United States was its monoculture of sugar, which tied the economy of the island to the oscillations and the speculations of the world market. I believe that this opinion was well founded. Yet the sugar monoculture has not disappeared under the Castro regime; it has continued, rather, to be the basis of the country's economy. "Socialism" has not enabled Cuba to change its economy: what has changed is its dependence.

Another fateful circumstance: Asia and Africa have become combat areas. Since modern war is transmigratory, it would not be beyond all possibility for the battleground to shift tomorrow to Central America. All the wars of this period have been peripheral ones, and in all of them the combination of three elements is perceptible: the interests of the Great Powers, the rivalries between the States of a given region, and these States' internal contentions. The traditional model is thus repeated. The most rapid review of the history of the great conquests—whether those of Caesar or of Cortés, of Alexander or of Clive—suffices to reveal the presence, again and again, of the same fundamental situations: the conqueror invariably takes advantage of the rivalries between States and their internal divisions. Sometimes the imperial power uses a satellite State to realize its aims: Vietnam in Southeast Asia, Cuba in Africa and Central America. The United States has also used the military regimes of Latin America and elsewhere as instruments of its policy. But it has not been able to count on such active and intelligent allies as the revolutionary elites that, in many countries, have taken over the movements for independence and national resurrection. In other cases, empires use the internal quarrels of various countries in order to intervene militarily, as in Afghanistan.

In the conflict between Israel and the Arab countries, although the skein from which their history is woven is more entangled still, the same factors enter the picture: the interests and ambitions of the Great Powers; the rivalries between the States of the region, a particularly complicated situation in that the Arab side contains not one or two but a number of tendencies; the internal divisions of each one of the protagonists—Jews and Palestinians in Israel, Christians and Moslems in Lebanon, Bedouins and Palestinians in Jordan, Shiites and Sunnites elsewhere, not to mention ethnic and religious minorities such as the Kurds. The conflict is a complicated knot of economic and political interests, national aspirations, religious passions, and individual ambitions. The accords between Egypt and Israel were a victory of realism as well as of Sadat's clearsightedness and courage. But it was an incomplete victory. There will be no peace in this region as long as, on the one hand, Israel does not recognize frankly and forthrightly that the Palestinians have a right to a national homeland and, on the other hand, the Arab States and the Palestinian leaders in particular refuse to accept, willingly and for all time, the existence of Israel.

Addenda

After the publication of the preceding pages (in January 1980), the threads became snarled in a bloody, inextricable tangle. The heinous murder of Sadat did not achieve the fanatics' aim: the negotiations between Egypt and Israel went on, and the Israeli troops evacuated Sinai. But the death of the president of Egypt has made the desolation of the current situation even more apparent: not one of the present leaders—Arab, Israeli, or Palestinian—is equal in stature to Sadat, possessed of sufficient boldness, vision, and generosity to repeat his gesture and hold out a hand to the enemy. Jews and Arabs are branches of the same trunk, not only because of their origins but because of their language, their religion, their history. If they managed to coexist in the past, why are they killing one another now? In this terrible struggle, stubbornness has become suicidal blindness. None of the contenders can win a definitive victory or wipe out the enemy. Jews and Palestinians are doomed to live side by side.

The people of the Holocaust has not been generous with the Palestinians, who have fled and taken refuge in Jordan (a State invented by British diplomacy). Once there, they provoked a civil war that ended in their being crushed by the Bedouins, their brothers in blood and religion. Fleeing once again, they found asylum in Lebanon, a country celebrated for its pleasant and easygoing way of life and for the peaceful and civilized coexistence enjoyed within it by Moslems and Christians. The Palestinians reorganized their guerrilla forces once again and turned Lebanon into a base of operations against Israel. At the same time, together with the Syrian occupation troops, they contributed decisively to the dismemberment of the country that had taken them in, and fought a bloody war with the Lebanese Christians, who had sought the aid of Israel.

After the accord with Egypt, the Israelis were able to turn their attention to this new front. Determined to put an end to the guerrilla forces, they invaded Lebanon, neutralized the Syrian troops, and, with the aid of the Lebanese Christian militias, defeated the Palestinian guerrillas. In the course of these operations, the Christian militias attacked a camp of Palestinian refugees and murdered more than a thousand defenseless people. It was a cruel act of vengeance, which turned the Lebanese Christians, who had been the victims of the Palestinians, into their killers, and the Israelis into indirect accessories. Thus things came full circle: the martyrs turned into executioners and made martyrs of their executioners.

The success of the Israeli operation in Lebanon was due not only to their military superiority. The Jews counted on the aid of the Christians and a large part of the Moslem population, who had wearied of the Syrian occupation and of the excesses of the Palestinians who had made Beirut their general headquarters. A new Lebanese government has been formed, and it is foreseeable that the foreign troops—Syrian and Israeli—will withdraw from Lebanon in the months to come. Hence the possibility arises that peace will return to the region, providing that all parties—Israelis, Jordanians, and Palestinians—agree to enter into negotiations. Those directly interested must reject the intervention of nations obedient to Moscow's prompting, who seek only to fan the flames. I am thinking of Syria, and above all of Libya, dominated by a demagogic tyrant who is both a religious fanatic and a patron of terrorists. On the other hand, the triumph of Israel could well prove to be a Pyrrhic victory if its leaders yield to the temptation to regard it as a definitive solution to the conflict. The solution cannot be military; it must be political, and founded on the one principle that will guarantee both peace and justice: the Palestinians, like the Jews, have the right to a homeland.[1]

1. Even as I was writing these words, unfolding events showed that I was too optimistic. Not only did the Syrians refuse to leave their positions in Lebanon; their military power has become a major factor. Externally, the Syrian regime has the military and political support of the Soviet Union. Internally, President Assad's power rests on a Shiite subsect called the Alawites. They are a minor-

It is true that the Palestinians' methods of fighting for this right have been, almost without exception, abominable; that their policy has been fanatical and intransigent; that their friends and defenders have been and are aggressive and criminal governments such as that of Libya and the totalitarian ideocracies. Yet, however grave all of this may have been and still is, it does not invalidate the legitimacy of their aspiration. Admittedly, they have followed fanatical leaders and demagogues who have led them to disaster: if they want independence and peace, they must seek out other leaders. But, at the same time, it must be conceded that Israeli intransigence, Jordanian self-interest, and the devious policy of a number of Arab States (Syria and Libya in particular) have been the other side of the coin of the demagogy and fanaticism of the Palestinian leaders. During World War II, André Breton wrote, "The world owes the Jewish people reparation." The moment I read them, I took these words to my heart. Forty years later I say: Israel owes the Palestinians reparation.

The battles and disasters of the Middle East are repeated and multiplied in Asia and Africa. In addition to the senseless war between Iran and Iraq, we have witnessed helplessly a succession of other conflicts: between Ethiopia and Eritrea, Libya and the Sudan, China and Vietnam, to cite only the best-known ones. These conflicts have ranged from border incidents to outright war, and from the killing of innocent parties to genocide. In all of them, interrelated in varying ways, the factors that I mentioned earlier have been present: the old tribal, national, and religious rivalries; the appearance of new dominant castes with authoritarian, aggressive ideologies, organized along military lines; and the armed intervention of the Great Powers. In

---

ity in a Sunnite country. Whatever the vitality of the Syrian regime, it is clear that Israeli policy suffered a serious setback. We can say the same of the United States and the other Western powers who had to withdraw their forces from Beirut. These failures confirm the sterility of a policy based solely on military considerations.

all these countries the fight against the domination of colonial powers took place under the leadership of an elite of professional revolutionaries. In many cases the leaders came by their revolutionary ideologies in the universities of the colonial mother countries, just as in the last century the Jesuit schools were the nurseries for atheists and freethinkers. Europe passed revolutionary ideas on to the leading classes of its colonies, and along with them the sickness that has been eating it away since the last century: nationalism. To the mixture of these two explosive elements—revolutionary utopias and nationalism—another, even more volatile one must be added: the emergence of a new elite whose ideas, organization, and fighting tactics duplicate the communist model of the party-as-militia. This elite consists of minorities trained and organized to wage war, steeped in aggressive ideologies, and not at all respectful of the opinions or even the existence of others.

Borne on the wave of genuine popular revolts, the elites of professional revolutionaries have allied themselves with the legitimate aspirations of their peoples only to pervert them. Once in power, they have instituted ideological war as the one and only *modus faciendi*. Marx believed that socialism would bring an end to war between nations; those who have usurped his name and his heritage have made of war the permanent condition of nations. Their action within their countries is despotic; outside their borders, invariably bellicose. The saddest and most terrible case has been that of Indochina: the defeat of the United States and of its allies resulted immediately in the installation of a military-bureaucratic regime throughout Vietnam. The communist government, violently nationalistic, revived long-standing Vietnamese pretensions to hegemony, and, aided and armed by the Soviet Union, imposed its domination by force of arms on Laos and Cambodia. The Chinese, following in turn their country's traditional policy in that region, have opposed the expansion of Vietnam and have succeeded in halting it but not in driving it out of Laos and Cambodia. In

luckless Cambodia, the Vietnamese troops and their local accomplices have overthrown the criminal tyrant Pol Pot, the
protégé and puppet of the Chinese. Pol Pot and his henchmen
were the authors of one of the great criminal operations of our
century, comparable to those of Hitler and Stalin. Nonetheless,
their downfall was not a liberation but the replacement of a
tyranny of pedantic assassins (Pol Pot studied in Paris) with a
despotic regime backed by foreign troops. The example of Indochina is striking because it shows us, with awesome clarity,
the fate that awaits popular uprisings taken over by elites of
professional revolutionaries organized along military lines. The
same process has been repeated in Cuba, Ethiopia, and today
in Nicaragua.

The Upsurge of Particularisms

For half a century there has been talk of the revolution of science and technology; other groups have spoken as insistently
of the revolution of the international proletariat. These two
revolutions represent, to ideologues and their believers, the two
contradictory yet complementary faces of the same deity:
Progress. From this point of view, returns to the past and historical resurrections are either unthinkable or reprehensible. I
do not deny that science and technology have radically altered
humanity's ways of life, though not its deeper nature or its
passions; nor do I deny that we have witnessed many upheavals and social changes, though their theater has not been the
countries predicted by doctrine (the developed countries), nor
its agents the industrial proletariat. However, what characterizes the last years of this century is the return of beliefs, ideas,
and movements that had supposedly vanished from the sur-
face of history. Many ghosts have come back to life; many
buried realities have reappeared.

If there is one word that defines these years, it is not "revolution" but "revolt." "Revolt," however, not only in the sense of disturbance or violent change from one condition to another but also in that of a turning back to origins—revolt as resurrection. Almost all the great social upheavals of recent years have been resurrections, the most notable being of religious feeling, generally allied with nationalist movements: the awakening of Islam; the religious fervor in Russia after more than half a century of antireligious propaganda, and the return, among the intellectual elite of that country, to modes of thought and philosophies that were believed to have disappeared with czarism; the revival of traditional Catholicism in the face of the conversion of part of the clergy to a revolutionary secular messianism (Mexico, Poland, Ireland); the wave of Christian revivalism among young people in the United States; the sudden popularity of Eastern cults; and so on. An ambiguous portent, since religions are what languages were to Aesop: the best and the worst thing ever invented by mankind. They have given us the Buddha and Saint Francis of Assisi; they have also given us Torquemada and the high priests of Huitzilopochtli.

The events in Iran are perfectly consistent with this conception of revolt as resurrection. The overthrow of the shah did not result in a victory of the liberal middle class, or of the communists: it was Shiism that won out. This outcome disconcerted everyone—the experts first of all, as has become the case more often than not. Shiism is more than a Moslem sect and less than a separate religion. Its faithful consider themselves to be the real orthodox followers of Islam and regard the practices and beliefs of the Sunnite majority, in their eyes very nearly heretics, as thoroughly contaminated with paganism. Shiism distinguishes itself from this majority by its puritanism, its intolerance, and the institution of the spiritual guide, the imam (among the Sunnites the imam is simply the believer entrusted with leading prayers in a mosque). The Shiite imam is a spiritual dignitary distinct from the caliph of the Sunnites. The caliphate was an office with features both of an elective pontificate and of a hereditary monarchy, whereas the imamate was a spiritual lineage. The imams, first of all, were direct descendants of Ali, the son-in-law of the Prophet, and of his grandson Husain, the martyr assassinated at Karbala; and they were the elect of God as well. The conjunction of these two circumstances, inheritance and divine election, underscores the theocratic nature of Shiism. There were twelve imams, and the last was the hidden imam, the One Who Has Disappeared, who will one day return: the Mahdi. This event, like the advent of Christ at the end of time, gives Shiism a metahistorical dimension.

Another notable feature is that all imams died violent deaths, though at the hands not of Christians or unbelievers but of Sunnite Moslems, victims of civil wars that were also religious wars. If there is one particular feature that distinguishes Shiism from the rest of Islam it is the cult of martyrdom: the eleven imams of tradition offered themselves in sacrifice, like Jesus. But there is a great and significant difference: all of them died with weapons in hand, or were poisoned by their enemies. Islam is a militant religion and a religion of warriors. What characterizes Shiism is that, unlike the Sunnism of the majority, it is a faith of the vanquished and of martyrs. In all religions, as in all manifestations of eroticism, there is both a sadistic side and a side inclined toward self-flagellation and martyrdom. In Shiism the second tendency wins out. However, as also happens in the realm of the erotic, the shift from masochism to sadism is a violent one that takes place with breathtaking suddenness. That is what has happened today in Iran.

A major difference between Shiites and other Moslems is the existence of an organized clergy, the guardian of not only religious but also national traditions. Shiism has been identified with the Persian tradition, and in certain of its sects—I am thinking of the Ismailis—the traces of the old religions of Persia, such as Manichaeism, are perceptible. It is to the point here to recall that the Persian genius has created great religious systems. In the Islamic period its great mystics were the glory of Sufism, which is the spiritual counterpart of Shiism. It is a people of philosophers, of visionaries, and of poets, but also of bloody prophets, such as Hassan Sabbah, the founder of the Hashshashin sect (the origin of the word "assassin"), hashish-smoking warriors whose secret acts of murder terrorized Christians and Sunnite Moslems in the twelfth century. To sum up: Shiism is a militant theocracy that leads to a metahistory: the cult of the Mahdi, the hidden imam. In its turn, Shiite meta-history results in a milleniarism at once nationalistic, religious, and combative, and fascinated by the cult of martyrdom.

The revolt that overthrew the shah and his regime is a trans-lation in more or less modern terms of all the elements that I have mentioned above. I emphasize, once again, that we have not witnessed a revolution in the modern sense of this word, whether liberal or Marxist, but of a *revolt*: a return to the in-nermost nature of the people, a bringing to light of the hidden tradition, a return to the original source. Iran has rejected the modernization from the top that the shah and his authoritar-ian regime, with the friendship and the aid of the United States, tried to impose upon it. On the downfall of Shah Mohammed Riza Pahlevi, many of us asked ourselves: Will the new leaders be able to conceive another plan for modernization, more in keeping with the country's tradition, and will they be able to carry it out from the bottom up? In the beginning there was good reason for doubt, although the presence in the Teheran government of such personalities as Bani-Sadr seemed to be a hopeful sign. This young political leader seemed for a short period to represent a bridge between the middle-class refor-mists and the followers of Ayatollah Khomeini, possessed by a politico-religious fury. Bani-Sadr came from a prominent reli-gious family; his father was an ayatollah, and it was Khomeini himself who presided at the latter's funeral. A theologian and an economist, Bani-Sadr proposed to bring about a synthesis of the Islamic tradition with modern political and economic thought. He did not even have time to formulate his ideas, however, swept aside by the hosts of his former friend and spiritual father, Khomeini. A deplorable episode: in the Persian revolt there were seeds of a historic renaissance.[2] Perhaps they still exist, though they lie submerged by the two figures that threaten any popular uprising: the demagogue and the tyrant. The demagogue provokes chaos; then the tyrant presents himself with his gallows and his hangmen.

Why, instead of opening doors, as the Mexican Revolution did, has the Iranian revolt closed them? In the movement against Mohammed Riza Pahlevi's regime the intervention of the enlightened middle class was decisive. This phenomenon is repeated again and again in history: the opposition to the shah was initiated by a social group that had emerged as a result of the policy of economic and intellectual modernization undertaken by that ruler himself. These intellectuals, receptive to modern culture and often educated abroad, were inclined toward a democratic nationalism tinged with socialist reformism. Certain of them, like Bani-Sadr, tried to reconcile modern thought with Islamic tradition (just as among us a number of Christian movements have attempted to bring together the old and the new). But this middle class, after ineffectually resisting Khomeini's bands, had no other recourse than to withdraw and yield power to the extremists. Many of them were executed and others forced into exile.

Khomeini's supporters are united by a traditional ideology, simple and powerful, that has identified itself, like Catholicism in Poland and in Mexico, with the nation itself. Faithful to the tradition of Islam—a religion of warriors—they organized themselves along military lines from the very start. Thus, in

---

2. "Iran," its official name, tends to underscore the Aryan origin of the nation. But "Persia" is a word that evokes three thousand years of history.

the Shiite bands loyal to Khomeini and his ayatollahs, we find
the same basic elements that distinguish communist parties:
the fusion between the military and the ideological. The content is different but the elements and their fusion are identical.
Their enormous economic interests threatened by the shah's
agrarian reform and their ideology by his policy of modernization, the Shiite clergy made common cause with the middle-
class reformists, but then within a very short time the clergy
took over the movement, which little by little turned into an
insurrection. In political terms, it was a revolt; in historical and
religious terms, a resurrection. Shiism had been a passive belief of the majority; it now became an active force in the political life of Iran. But because of both its ideology and its vision
of the world, and because of its structure and organization, the
Shiite clergy cannot become the lever that will prize open the
doors leading to genuine modernity for the Iranian people. The
movement is purely and simply a regressive one. A cruel dis-
appointment: the revolt ended in frenzied clerical domination,
the resurrection in a relapse.

Tyrannies and despotisms need the threat of an outside enemy to justify their rule. When such an enemy does not exist,
they invent one. The enemy is the devil. Not just any devil,
but a figure, half real and half mythical, in which the enemy
without and the enemy within are conjoined. The identification of the domestic enemy with the foreign power possesses,
at one and the same time, a practical as well as a symbolic
value. The devil is no longer within us but outside the social
body: it is alien, and we must all rally round the revolutionary
chief to defend ourselves. In the case of Iran, the devil Carter
was the agent of revolutionary unity. It was imperative for
Khomeini to achieve this unity. Without the devil, without the
foreign enemy, it would not have been easy for him to justify
the fight against the ethnic and religious minorities—Turks,
Kurds, Baluchi, Sunnites—and against nonconformists and
dissidents.

I do not mean to say that the Iranians' anger against the United States was unjustified. To the Moslem mind, the Americans represent the continuity of Western domination; they are the heirs not only of the European imperialism of the past century but also of the adventurers and warriors of other centuries. In addition to these historical obsessions, a contemporary reality also played a part: succeeding administrations in Washington were more than just friends of the shah, they were his accomplices and defenders. Hence everything conspired to make the United States the inevitable devil of the Iranians. It cannot be said that that nation had done nothing to deserve that dubious distinction. The arrival of the shah in New York triggered the fusion between imagination and reality: the devil ceased to be a mere notion and, to believers, suddenly turned into a palpable presence. The response was the assault on the American embassy and the capture of the American diplomats.

The episode has the air of a piece for the theater with a plot woven by Chance, an author more indifferent than evil: the same author who in Shakespeare's and Marlowe's works takes the place of Greek Destiny and Christian Providence. The difference between these age-old powers and modern Chance is this: it is presumed that the acts of Providence and of Fate have a meaning, though it may be a hidden one, whereas those of Chance have no logic, purpose, or meaning. Each one of the elements of the crisis was linked to the next with a sort of rigorous incoherence and a total lack of premeditation: the shah's illness, his imprudent decision to undergo treatment not in Mexico but in New York, the no less imprudent decision on the part of the authorities in Washington to allow him to enter the country. In the eyes of the Iranian leaders, the presence of the shah in New York was a gift, not fallen from heaven but handed to them by their adversaries themselves.

The attack on the embassy and the seizure of the diplomats as hostages was a sort of sacrilege. In revolutionary movements the notion of sacrifice is invariably closely linked with that of sacrilege. The victim symbolizes the order that is dying and his blood suckles the time that is aborning: Charles I is beheaded and Louis XVI is guillotined. The function of the sacrilegious act is similar to that of the foreign devil: it unites the revolutionaries in the brotherhood of shed blood. Balzac was one of the first to show how shared crime is a sort of communion (*Histoire des Treize*). Sacrilege, furthermore, desacralizes the person or the institution profaned; by that I mean that it is a profanation in the true sense of the word: what was sacred becomes profane. To break into and occupy the United States Embassy was to profane a place traditionally considered by treaties, international law, and custom to be inviolable. Such a profanation simultaneously proclaims that there is a higher law: revolutionary law. This process of reasoning is not juridical but religious: revolts and revolutions are myths incarnated.

In Iran, sacrifices have been and are numerous, though they have not had the impersonal character of the mass murders of Hitler, Stalin, and Pol Pot, who applied the efficient methods of industrial assembly lines to the process of exterminating their fellow humans. The cruelty of Khomeini and his clergy is archaic. In sacrifice, as in the rite of the foreign devil, political usefulness and ritual symbolism go hand in hand. All revolutionary movements claim to found a new order or to restore an immemorial one. In both cases revolutions, true to the original meaning of the word, are returns to the beginning, a real starting all over again. The act of beginning anew, as anthropology teaches us, is actualized or realized through a sacrifice. Between the time that is ending and the time that is beginning there is a stop: sacrifice is the act whereby time is set in motion again. This is a universal phenomenon, present in all societies and eras, though in each one it assumes a different form. In certain isolated regions of India, for example, the building of a house begins, even today, with a rite that consists of wetting the foundations with the blood of a baby goat, a substitute for what used to be a human victim. A curious transposition of this old rite into modern political terms: some years ago, in the course of a diplomatic visit I paid Mrs. Bandaranaike, at the time the prime minister of Ceylon (today Sri Lanka), she said to me, commenting on one of her trips to Peking, "The Chinese have an advantage over us in that they really had to fight. It was unfortunate for Ceylon that we obtained our independence without an armed struggle and almost without bloodshed. In history, in order to build it is necessary to wet the bricks with blood. . . ."

In the incident of the hostages, the liturgy was observed only symbolically—although there was profanation and sacrilege, sacrifice was not made. Nor was there a public trial: the Iranian government did not carry out its threat to judge the diplomats and punish those found guilty. This, again, is a liturgical act. Since revolutions claim to restore the just order of the beginning, it is necessary for them to have recourse to procedures that will turn the inviolable person (king, priest, diplomat) into an ordinary individual, and the victim into a criminal. Primitive societies resort to magic to change the nature of the victim; modern ones, to criminal justice. This was the reason for the trials of Charles I, Louis XVI, and in our time those of Bukharin, Radek, Zinoviev, and the other Bolsheviks.

The Khomeini regime has transformed the conflicts with its neighbors into an ideological war and a religious crusade. It has thus followed the inevitable course of all revolutions and been faithful to the Shiite tradition of a holy war against its Sunnite brothers. In 1980, in the first version of this essay, I wrote:

Shiism is belligerent, and just as it has provoked violent rejection on the part of the ethnic and religious minorities of Iran, it must necessarily confront the other Moslem countries of the region sooner or later. For the Shiite clergy, as for the imams of the past, religion, politics, and war are one and the same thing. Hence, both out of fidelity to its tradition and out of obedience to the necessities of the present, it will attempt to restore the endemic state of war that for centuries character- ized the Islamic world. The governments of Iraq, Syria, and Saudi Arabia know this—even better than Washington and Moscow.

The war with Iraq bore out my predictions.

Iraq is a country in which military dictatorships, armed by the Soviet Union (and, alas, by some Western powers as well), have followed one on the heels of another for years, disguised as Pan-Arab socialism. At the beginning of the hostilities, the "experts" were certain that Iraq would soon triumph over a nation divided and bloodied. Thus far, however, Iran has been the victor. The reason for its success in battle is not so much military and political as religious: its armies are possessed by a faith. Sacrifice of self, once again, in the service of a perversion.

The most notable feature of the conflict between the United States and Iran, the one that makes of it a classic example, is the inability of either side to understand what the other was saying. It was impossible for Khomeini to understand the Americans' juridical and diplomatic arguments. He was pos- sessed by a religious and revolutionary fury—the two adjec- tives are not contradictory—and Carter's language must have struck him as secular and profane—that is, satanic, inspired by the devil, the father of falsehood.

Nor could the Americans understand what Khomeini and his followers were saying: it struck them as the language of madmen. Again and again they dismissed the ayatollah's state- ments as incoherent, delirious ranting. For the modern con- sciousness, irrationality and delirium are what demoniacal possession was for the ancients. Hence there is a certain sym- metry between the attitude of the Americans and that of the Iranians. Carter was possessed by the devil, or, in other words, he was mad; Khomeini was delirious, or, in other words, he was possessed by the Evil One. To put it another way: Khomeini's language is that of other centuries and the Americans' language is modern. It is the optimistic, rationalist language of liberalism and pragmatism, the language of the bourgeois democracies, proud of the conquests of the physical and natural sciences that have given them dominion over nature and over other civilizations. But neither science nor technology can save us from natural or historical catastrophes. Americans and Europeans must learn to hear the *other* language, the buried language. Khomeini's language is archaic, and at the same time it is profoundly modern: it is the language of a resurrection. Learning this language means rediscovering that wisdom forgotten by modern democracies but never by the Greeks, save for those times when, out of sheer exhaustion, they forgot themselves: man's tragic dimension. Resurrections are terrible; while today's political leaders and heads of government may not have learned this yet, poets have always discovered that secret. Yeats did:

> . . . somewhere, in sands of the desert
> A shape with lion body and the head of a man,
> A gaze blank and pitiless as the sun,
> Is moving its slow thighs, while all about it
> Reel shadows of the indignant desert birds. . . .

V

Mutations

Grafts and Rebirths

I have discussed the case of Iran at length because it seems to me that it is a sign of our times. The resurrection of national and religious traditions is merely one more manifestation of what can only be called the historical vengeance of particularisms. This has been the dominant theme of these years and will be that of the years to come. Blacks, Chicanos, women, Basques, Bretons, Irishmen, Walloons, Ukrainians, Latvians, Lithuanians, Estonians, Tatars, Armenians, Czechs, Croats, Mexican and Polish Catholics, Buddhists, Tibetans, Shiites of Iran and Iraq, Jews, Palestinians, Kurds repeatedly murdered, Lebanese Christians, Marathas, Tamils, Khmers—each of these names designates an ethnic, religious, cultural, linguistic, sexual particularity; all of them are irreducible realities that no abstraction can dissolve. We are experiencing the rebellion of exceptions, no longer suffered as anomalies or violations of a supposed universal rule, but assumed as a unique truth, as a destiny. Marxism had postulated a universal category—classes—and used it not only to try to explain past history but to make future history as well: the bourgeoisie had made the modern world, and the international proletariat would make the world of tomorrow. Positivism and liberal thought reduced the plural history of men to a unilinear and impersonal process: progress, the offspring of science and technology. All these conceptions were tinged with ethnocentrism, and both old and new particularisms have taken up arms against them. The pretended universality of the systems elaborated in the West during the nineteenth century has been shattered. Another universalism, a plural one, is dawning.

The resurrection of peoples of old and their cultures and religions would have been impossible without the presence of the West and the influence of its ideas and institutions. European modernity was the reagent setting off the tremors that shook the societies of Asia and Africa. The phenomenon is not new: history is made up of impositions, borrowings, adoptions, and transformations of alien religions, techniques, and philosophies. Contacts with outsiders are a decisive factor in social change. This has been particularly true in Asia and in northern Africa, the seats of old cultures: in China it is no longer the Son of Heaven who governs but the secretary general of the Communist Party; Japan is a democratic monarchy; India is a republic; Egypt is yet another.

Regimes, ideologies, and national flags have changed, granted, but have the deeper realities changed as well? If, for example, we subject the area of relations between nations and States to close scrutiny, what do we find? We have only to reread a work of fiction such as Kipling's *Kim* (1901) to realize that the broad canvas that serves as a historical backdrop for the novel's plot is not all that different from the situation today: the struggles, over a vast region stretching from Afghanistan to the Himalayas, between the imperial ambitions of the Russians and those of the Western powers. The rivalry between the Chinese and the Russians began in the seventeeth century. The relations between China and Tibet have been unstable and turbulent since the thirteenth century: armed conflicts, occupations, rebellions. The enmity between Chinese and Vietnamese began in the first century B.C. The history of Cambodia has been, since the fourteenth century, one constant battle with its two neighbors, Thailand and Vietnam. And so on. Is there, then, nothing new under the sun? On the contrary: the difference between the Asia of 1880 and that of 1980 is enormous. A century ago the Asiatic countries were the theater of the struggles and ambition of the European powers; today these old peoples have awakened, have ceased to be objects, and become subjects of history.

The first forewarning of the change occurred in 1904, when the Japanese defeated the Russians. Since then the phenomenon has manifested itself in various political and ideological forms, from Gandhi's nonviolence to Mao's communism, from Japanese democracy to Khomeini's Islamic republic. The great mutation of the twentieth century has not been the revolution of the proletariat in the industrial countries of the West but the resurrection of civilizations that gave every appearance of having turned to stone: Japan, China, India, Iran, the Arab world. On the brutal but vivifying contact with European imperialism, they opened their eyes, rose again from the dust, and began to stir. Today these nations are confronting a similar problem, one that each of them is endeavoring to resolve in its own way: modernization. The first to have succeeded is Japan. What is most significant is that its version of modernity did not destroy its traditional culture. The error of the shah, by contrast, was to try to modernize from the top down, trampling popular customs and sentiments underfoot. Modernization does not mean mechanically copying the United States and Europe: to modernize is to adopt and to adapt, but it is also to re-create.

The success of the Japanese has been truly exceptional: in 1868, at the beginning of the Meiji era, they decided to modernize, and half a century later they were already a great power, economically and militarily. The most difficult modernization, that of their political life, was achieved more slowly, and not without relapses. During this process—which took approximately a century—Japan contracted the three diseases of modern Western societies: nationalism, militarism, and imperialism. After its defeat in World War II, and after having been the victim of the criminal American attacks on Hiroshima and Nagasaki, the Japanese rebuilt their country and in the process turned it into a modern democracy.

The Japanese experience is unique, not only because of the rapidity with which they assimilated and made their own the sciences, techniques, and institutions of the West but also because of the originality and inventiveness with which they adapted them to their country's particular genius. The transition from one era to another provoked, as goes without saying, conflicts and rents both in the texture of society and in the innermost hearts of individuals. Novels, essays, and social and psychological studies dealing with this subject abound; nonetheless, though there is no denying that these social changes and this turmoil were profound, Japan did not lose its cohesion, nor did the Japanese lose their identity. Within Japanese tradition, moreover, there are other examples of borrowings from abroad that were assimilated and recast with the greatest felicity by the native genius. In other words, Japanese tradition, from the seventh century A.D. on, has been a unified whole of foreign ideas, techniques, and institutions, mainly Chinese (writing, Buddhism, the moral and political thought of Confucius and his heirs), altered and transformed into new and authentically Japanese creations. The history of Japan confirms Aristotle: all true creation begins as imitation.

At the other extreme is India. Unlike Japan, which is a single, indivisible, homogeneously constituted nation, India is a whole constituted of heterogeneous peoples, each with its own language, tradition, and culture. Before the British raj, there had been no such thing as a truly national State in India; the great empires of the past—the Maurya, the Gupta, the Mongol—did not rule over the entire subcontinent, nor were they really national. In the languages of northern India there is no word with the meaning of the historical reality that in Western languages denotes the concept of nation. The unity of the peoples of India was not political but religious and social: Hinduism and the caste system. Thus it has been said that India is not a nation but a civilization. The foundation of Indian society is Indo-European. This is the essential and decisive fact: Hinduism, that totality of beliefs and practices in which Indians have been steeped for more than three thousand years, and which has given them unity and an awareness of belonging to a vaster community than their particular nations, is of Indo-Aryan origin. The language of the sacred and of philosophy is likewise Indo-European: Sanskrit. So is the caste system, which is a modification of the tripartite division of the religion, thought, culture, and society of the ancient Indo-Europeans, as Georges Dumézil has shown with brilliance, profundity, and immense erudition. I emphasize that the quadripartite division of Indian castes is not a change but a modification—through addition—of the original Aryan system. For all of these reasons it can be said that India, lying between the Far East and Western Asia, is "the other pole of the West, the *other* version of the Indo-European world"—or more exactly, its inverted image.[1]

---

1. I have touched upon the subject of the symmetrical opposition between the West and the civilization of India in *Alternating Current* and in *Conjunctions and Disjunctions* (1974; original Spanish edition, *Conjunciones y disyunciones*, 1969). In these two books I have also explored to some extent the similarities and antagonisms between traditional Indian thought and that of the West (the notions of being, substance, time, identity, change, etc.). I have dealt with the caste system in *Claude Lévi-Strauss* (1973; original Spanish editon, *Claude Lévi-Strauss o el nuevo festín de Esopo*, 1967), as well as in the two books cited above and in a rather long note in *Children of the Mire* (1974; original Spanish edition, *Los hijos del limo*, 1974). As for Hinduism, although it descends, directly and essentially, from the ancient Vedic religion, which was Indo-European, I am not unaware that it bears traces of Dravidian beliefs, such as worship of the Great Goddess and of a proto-Siva.

I must add that the definition of India not as a nation but as a civilization is misleading. Actually, two civilizations live and struggle side by side in the subcontinent: the Hindu and the Islamic, not to mention the primitive cultures and tribes and the mixed cultures, such as the Sikhs, who owe so much to both the Islamic and Hindu traditions.

The originality of India as compared with the other two great Indo-European communities—the Iranian and the European properly speaking—is twofold. On the one hand, many of the original institutions and ideas of the Indo-Europeans appear in India almost intact, with a sort of immobility that, though it may not be that of death, is at the same time not that of real life, either. On the other hand, unlike Europeans and Iranians, polytheistic India has, for eight centuries, found coexistence with a severe and intransigent monotheism extremely difficult. In Europe, Christianity succeeded in synthesizing the old Indo-European paganism—with its gods, and its vision of being and of the universe as self-sufficient realities—and Jewish monotheism and its idea of a Creator God. Without the synthesis of Christian Catholicism and the subsequent criticism of this synthesis, begun in the Renaissance and the Reformation and continued on through the eighteenth century, the prodigious achievements of the West would not have been possible. We owe to critical European thought the gradual introduction of the notions of history as successive change and as progress, ideological assumptions underlying the action of the West in the Modern Age. In Iran, too, Semitic monotheism triumphed, though in its Islamic version. The Iranian synthesis was less fecund than the Christian, both because of the exclusivist nature of Islam and because the Iranian substratum was less rich and complex than the Greco-Roman. The ancient Iranian tradition was only partially adopted by Islam; the rest was buried and repressed. Iran had no movement comparable to the European Renaissance, which was at once a return to pagan antiquity and the beginning of modernity.

Unlike Europe's, India's experience included neither the idea of history nor that of change; by this I mean that though changes did certainly occur, India neither reflected upon them nor interiorized them. Its vocation was religion and metaphysics, not historical action or the rule of natural forces. Hence, by contrast with Iran, in India Hinduism has coexisted with Islamic monotheism without really living on intimate terms with it: it has neither been converted to Islamic faith nor been able to absorb it. This is the root, in my opinion, of the divergent paths followed in their evolution by India on the one hand and by the other two great Indo-European areas on the other.

The result is apparent at first glance: in 1984 India had nearly 725 million inhabitants, of whom 10 to 12 percent were Moslems—in other words, something like eighty million. But that figure is misleading, since properly speaking there should also be included within it the population of the two countries that have separated from India but nonetheless still have a language, a culture, and a history that are Indian: the nearly one hundred million people of Bangladesh (85 percent Moslem) and the more than ninety million people of Pakistan (nearly all Moslem). This would make in all more than 250 million Moslems—an enormous number.[2] The independence of the Indian subcontinent coincided with the bloody secession of Pakistan; the division was due, fundamentally, to the fact that Hindus and Moslems alike found it impossible to conceive of living together. There were terrible massacres, and entire populations were removed from one territory to the other. The historical cause of this disaster—a wound that has not yet healed—is the one indicated above: in India there was never a synthesis such as that brought about by Christian Catholicism in Europe, nor was there an absorption of one religion by another, as in Iran.

Since 1947 the foreign policy of India has been primarily determined by its obsessive preoccupation with its politico-religious quarrel with Pakistan. The same can be said of the Pakistanis. But is it really a matter of *foreign policy*? The rivalry was born at a time when neither the Indian State nor the Pakistani State yet existed; it was a religious and political contest

2. These are approximate figures, taken from the 1984 edition of the *World Book Encylopedia*.

between two communities within one and the same society, communities that spoke the same language, shared the same land and the same culture. It is no exaggeration to say that the conflict between India and Pakistan has been and is a civil war that began as a religious war. The occupation of a part of Kashmir by India, Pakistan's friendship with the United States first and later on with China, India's friendly gestures toward Russia, its deviations from its policy of neutrality, and its support of a number of deplorable resolutions within the Non-aligned Movement, as well as its hypocritical closing of its eyes to the occupation of Cambodia by Vietnamese troops—all the foregoing are in reality simply the struggle between two politico-religious factions. Doubtless it is too late now to unite what was separated; but it is not too late to create a sort of federation of India, Pakistan, and Bangladesh that will guarantee the peaceful coexistence of the two communities. Like the bitter struggle between Arabs and Jews, that between Indians and Pakistanis disproves, yet again, the supposed rationality of history.

Despite its traditionalism, India, too, has been capable of assimilating and transforming ideas and institutions that come from outside. Gandhi's movement, which was at once spiritual and political, was one of the great historical novelties of the twentieth century. Its origins are intimately related to the history of the Indian Congress Party, a group that came into being as a consequence of the democratic ideas brought to the subcontinent by England, and also owed much to a Scottish theosophist (A. O. Hume) who played a decisive role in its founding in 1885. But Gandhi's political action is inseparable from his religious ideas, in which we find an impressive combination of Hindu and European elements. The foundation was Hindu spiritualism, above all the *Bhagavad Gita*; alongside it, passed on by his mother, was the Vishnuism of his childhood, steeped in Jainism (the source of his teachings on nonviolence toward every living creature: called *ahimsa*); plus Tolstoyan

Christianity; and Fabian socialism. The essential contribution was Hinduism, though it is significant that Gandhi first read the *Gita* in Arnold's English translation. It is significant, too, that Gandhi's murderer was a member of a fanatical group also primarily inspired by the *Gita*. Different readings of the very same text . . .[3] Another characteristic of Gandhi's odd relationship with Hindu tradition was that, although his ideas and his ascetic practices were those of a true *sannyasi*,[4] and in his autobiography he says, "What I have sought and yet seek is to see God face to face," he sought God not in the solitude of the hermit's cave, hidden away from the world, but amid the multitudes and in political discussions. He sought the absolute in the relative, God among men, thereby uniting Hindu and Christian tradition.

Rejecting, on the one hand, the tactics and techniques of the political leaders of the West, based almost invariably on propaganda tricks, and, on the other, the strategy of the proponents of violence (everything that furthers revolution, Trotsky said, is good and moral), Gandhi adopted a new type of action: *satyagraha*, firmness in the truth and nonviolence. This policy has sometimes been criticized as impossibly idealistic, sometimes as hypocritical. Both Marxists and rightist realists and cynics have, as one, heaped sarcasm and bitter invective on Gandhi and his disciples. There is, nonetheless, no denying that Gandhi was able to stir immense multitudes. Consider the recent and well-known instance of Martin Luther King, whose movement against racial discrimination, inspired by Gandhi's methods, jolted and moved the whole of the United States. No less a figure than Einstein thought that only a universal movement that would carry on Gandhi's lesson of nonviolence could pressure the Great Powers into renouncing the use of nuclear

3. Krishna's discourse to dissipate Arjuna's doubts and fears before the oncoming battle is a vision of war as the fulfillment of the *dharma* (law, proper way) of the warrior.
4. A holy man who abandons the world. (TRANS.)

weapons. It will be said, not without reason, that Gandhi's movement was able to gather force and prosper only because the British government, imperialist or not, respected basic freedoms and human rights. Would a Gandhi have been possible in Hitler's Germany? In our own time, today, could a Gandhi arise in China, in Russia, in South Africa, in Paraguay? All that must be granted, but it must, equally, be granted that Gandhism is the one movement that has been able to offer a civilized and *effective* answer to the universal violence unleashed in our century by dictators and ideologies. Like the libertarian tradition, it is a seed of salvation. The ultimate fate of both traditions is intimately linked to that of democracy.

Two other great political achievements of modern India are, first, the preservation of the national State, and, second, the continuance of democracy. Both institutions, transplanted by the British to the subcontinent, were adapted by the Indians to the particular genius of their country. Thanks to the first, India is now a nation; thanks to the second, it has confuted all those who see in the democratic system a mere excrescence of liberal capitalism.

I have seen the Indian multitude, poor and illiterate, go to the polls; it is a spectacle that renews man's hope—the precise contrary of the spectacle of the multitudes screaming and shouting in the stadiums of the West and of Latin America. Of course, the political democracy of India stands in sharp contrast to the poverty of its people and the country's terrible social inequalities. Many wonder whether it is not now too late to abolish such wretchedness: the demographic growth of India would appear to have definitely tipped the balance against this eventuality. I am not yet willing to believe it: the people who gave us the Buddha and Gandhi, who discovered the concept of zero in mathematics and nonviolence in politics, can find its own way to economic development and social justice. But if it should meet with failure, its defeat would be the foreshadowing of that of other countries, such as ours, which have not been able to attain an economic growth to counter-balance their population growth.

British rule gave India, for the first time in her history, institutions that encompass her different nationalities and cultures: a civil service, an army, and a juridical system for all (previously, each community was governed by its own particular laws and regulations). Another common institution was added with the rise of a supranational intellectual class, educated in England or at Indian universities, where they could assimilate European culture. This class acted as a guide and inspiration in the struggle for independence. The Indian state, as successor to the British Raj, has been the inheritor of these institutions, even if the basis of its legitimacy is different—the democratic consensus of the various peoples of India. Is it possible to call the Indian state *national*? This question gains dramatic force in the light of the bloody evolution of the Sikh movement—the criminal terrorism of the extremists, the violent occupation of the Golden Temple with its thousands of deaths, the murder of Indira Gandhi and in its wake the killings of Sikhs in Delhi and other cities by mobs spurred on by demagogues. The Indian State is and must be, like the British Raj, a supranational state, not resting on domination like its predecessor but on the concerns of different communities. Nehru's legacy and the legacy of Indira Gandhi consists primarily in preserving this consensus. If the processes of division continue, India will be transformed into another battleground of the Great Powers. To preserve her unity is to preserve peace.

The ancient Chinese called their country the Middle Kingdom. They were right to so name it. Although China is "in the middle" in neither a geographical nor a political sense, it is nonetheless what is known as a "central" country, in the sense of holding a key place. Its influence is already decisive in our day and will become increasingly so. China has been China for more than three thousand years: a territorial, ethnic, cultural, and political continuity. China has suffered invasions and occupations—by Mongols, by Manchus, by Japanese. It has known periods of splendor and others of decadence; it has experienced violent social changes. And yet it has never ceased to be a State. I will offer a few examples that illustrate this admirable continuity. The first is linguistic. In the philosophical and political vocabulary of China, there was no such concept as revolution, in the sense that this term has had in the West since the French Revolution: the violent replacement of one system by another. The concept in Chinese that was closest to revolution in this sense was Kuo Ming. But Kuo Ming really means "change of name" or "change of mandate," which by extension meant a change of dynasty or of the ruling house (Change of Heaven's Mandate). In the early years of this century, the great republican leader Sun Yat-sen decided to use Kuo Ming as a synonym for revolution, and thus there was born the Kuomintang, the party that was later to be overthrown by the communists. Thus the very expression that designates the concept of revolutionary change is steeped in traditionalism.

Another example: Mao Tse-tung. He does not resemble any of the revolutionary figures of the West: neither Oliver Cromwell nor Robespierre nor Lenin nor Trotsky; he resembles Shih Huang Ti, called the First Emperor, with whom, at the end of the second century B.C., one era ended and another began. So with Mao, centuries later, one period ended and another began. The work of the First Emperor was continued by his successors, though stripped of its radicalism, adapted to reality, humanized. At the same time, Shih Huang Ti's memory was execrated. From what we have seen in the last few years, the figure of Mao and his achievements are already beginning to suffer the same fate. A third example: the social group that ruled the empire for two thousand years, the mandarins, bears more than one resemblance to the group that rules China today: the Communist Party. The mandarins did not constitute a bureaucracy of technocrats, specialists in economics, industry, commerce, or agriculture: they were, rather, experts in the art of politics who tried to put the political philosophy of Confucius into practice. The Chinese communists are also experts in political affairs. The content changes; form and function remain. A curious contradiction: by attacking Confucius, the communists confirmed the continuity of the Confucian tradition.

The Cultural Revolution is another example of the intimate alliance of change and continuity. The criticism unleashed by Mao against the communist bureaucracy during the Cultural Revolution has no antecedents in the history of Marxist parties in the West. On the other hand, it is strikingly reminiscent of the philosophical and political anarchism of the intellectual current that was the rival of Confucianism: Taoism. China has periodically been shaken by popular revolts, almost always inspired by the libertarian spirit of Taoism and directed against the mandarin class and the Confucian tradition. By its violent attack on formalist culture and bureaucracy, as well as by the high value it placed on popular spontaneity, the Cultural Revolution may be seen as a recrudescence of the Taoist temperament in modern China. The Red Guards bear an odd resemblance to the Yellow Turbans of the second century or the Red Turbans of the fourteenth.

It is difficult to know with absolute certainty what motivated Mao to unleash the "Cultural Revolution." His power struggle with President Liu Shao-chi and the other communist leaders, who after the failure of his economic policy tried to shunt him aside, was doubtless a key factor. Mao's answer to their move was to open the floodgates of repressed popular anger—a strange spectacle that, once again, disproved both the arguments of Marxism-Leninism and the speculations of Western experts: an old man heading a rebellion of young people, a Marxist-Leninist launching an attack on the most perfect expression of the doctrine of the party as the "vanguard of the proletariat," the Central Committee and its functionaries.

The Cultural Revolution shook China to its foundations because it corresponded, at one and the same time, to the contemporary aspirations of Chinese society and to its libertarian tradition. It was not a revolution in the modern Western sense, nor, I repeat, was it a revolt. The movement went beyond what Mao had foreseen and came close to submerging the regime beneath wave after wave of anarchy. In order to bring the Red Guards under control, Mao was forced to backtrack and call upon the aid of Lin Piao and the army. Then, in order to get rid of Lin Piao, he was obliged to ally himself with the moderate wing and call upon Chou En-lai. This zigzag policy reveals that Mao was not so much the Great Helmsman as a clever and devious politician. He managed to remain in power, but the cost was enormous, and all these upheavals not only led nowhere but had extremely harmful effects. The worship of Mao was a tremendous setback to the intellectual and political life of China; the work of the last years of his life has all the terrifying unreality of a paranoid nightmare. In his lifetime he was showered with delirious praise and compared to the greatest names in history. We know now that he was not Alexander; he was most assuredly not Marcus Aurelius, or even Augustus. His figure is already enshrined in the monster gallery of history.

Under Deng Xiaoping's leadership, the Chinese government has undertaken the process of demolishing the idol. Like the successors of Shih Huang Ti, the group in power is confronted with a twofold task: changing the regime and at the same time keeping it in power. Mao's modernization was a cruel fantasy; Deng, a more sensible man, has proposed to carry out Chou En-lai's four modernizations: in agriculture, industry, defense, and science and technology. China has the natural resources, the population, and the political will to transform itself into a modern nation. Throughout their history the Chinese have given proof of great technical and scientific abilities: they were the discoverers of gunpowder, the compass, and printing. Although this tradition remained at a standstill for several centuries, due to adverse historical circumstances, Chinese creative genius never died out. The Chinese are an industrious, patient, responsible, hard-working people.

Modernization means adoption and adaptation of the civilization of the West, above all its science and its technology. The Chinese built an original civilization, founded on principles very different from European ones. Nonetheless, traditional Chinese civilization will not be an obstacle to its modernization. A few months ago *The Economist* of London pointed out that all the countries influenced by China—that is, those informed by the political and moral thought of Confucius—have modernized themselves more rapidly than the Islamic countries and many Catholic ones: Japan, Taiwan, Singapore, South Korea, and Hong Kong. And *The Economist* adds, "If the four modernizations are achieved, the miracles of South Korea and Singapore will seem like sunspots compared to the bright sun of China."[5] The contribution of the United States and Western Europe to the modernization of this immense country will be a key factor. Moreover, there is a present fact destined to change not just the history of Asia but of the world: the collaboration between the Japanese and the Chinese. In the early years of this century, in a work of fiction that was politically and philosophically prophetic, Vladimir Solovyev foresaw a collaboration between Japanese technology and Chinese manpower. The fantasy of the Russian philosopher will in all probability become a reality by the end of this century.

The Chinese past, I am certain, will not be an obstacle to its modernization, just as the past has not held back Japan and other countries marked by Chinese influence. And communism? In Russia its effects have been contradictory: it has

5. In the light of these experiences, the essay that Max Weber devoted to Confucianism and Taoism in relation to modernity is worth rereading—and revising.

brought about the industrialization of the country, but in other respects it has caused the country to retrogress. The Sinologue Simon Leys, the author of keen-sighted essays on Mao and his regime, thinks that the Chinese will be capable of doing with Marxism what they previously did with Buddhism: assimilate it, change it, and turn it into a creation of their very own. Why not? The Chinese genius is pragmatic, imaginative, flexible, and not at all dogmatic; in the past it achieved a synthesis between the puritanism of Confucius and the poetic anarchism of Lao-tzu and Chuang-tzu; tomorrow will it give the world a less inhuman version of communism? If so the government of Peking will be obliged to undertake forthwith the "fifth modernization," the one sought by the dissident Wei Jin-sheng: democratization.

What we call "modernity" was born with democracy. Without democracy there would be no science, technology, industry, capitalism, working class, or middle class, which is to say that there would be no modernity. It is quite true that without democracy a great political and military machine, such as that in Russia, can be built. But even aside from the fact that the social cost that the Russian people has had to pay has been extremely high and painful, modernization without democracy contributes technological innovations to societies yet does not change them. In other words, it turns them into stratified societies, into hierarchical caste societies.

The case of China presents particular problems because it has never had anything like democracy in all of its history. For millennia it was governed by a dual system: on the one hand, the emperor, with his court and the army; on the other, the mandarin caste. The alliance between the throne and the mandarins was unstable, broken repeatedly but repeatedly renewed. Although the communist revolution changed many things, the dual system continued: on the one hand, Mao—that is, the emperor and his retinue; on the other, the Communist Party—the reincarnation of the mandarins of old. An alliance no less unstable than the traditional one, as the Cultural Revolution proved. Do these precedents mean that China is not suited for democracy? No, I believe that democracy is a universal political form that can be adopted by all peoples, as long as each of them adapts the form to its own genius. If China were to turn toward freedom, it would usher in a new era in modern history. Admittedly, neither the ideology nor the interests of the dominant groups can give us much reason to hope that the regime will undertake the most arduous modernization of all but the only one worth the cost: moral and political modernization. I hope so nonetheless: my love and my admiration for the thought, the poetry, and the art of this country are stronger than my skepticism.

The Latin American Perspective

No Latin American can be unaware of the process of modernization in Asia and Africa. The history of our countries since Independence is the history of various attempts at modernization. Unlike the Japanese and Chinese, whose leap toward modernity has been made from non-Western traditions, we for our part are descendants—culturally and historically, if not racially—of a branch of civilization where the conjuncture of attitudes, techniques, and institutions that we call modernity first took place. It must be added, however, that we are descended from the cultures of Spain and Portugal, which isolated themselves from the general European current just as modernity was beginning. During the nineteenth and twentieth centuries, the Latin American continent has adopted successive schemes of modernization, all of them inspired by the American and European example, though to date none of our countries can, strictly speaking, be called "modern." This is true not only of those nations where the Indian past is still alive—Mexico, Guatemala, Peru, Ecuador, Bolivia—but also of those that are almost entirely European by origin, such as Argentina, Uruguay, and Chile. Moreover, Spain and Portugal are not fully modern, either. In our countries the burro and the airplane, illiterates and avant-garde poets, straw huts and steel mills coexist. All these contradictions culminate in one: our constitutions are democratic, but the real and omnipresent reality is dictatorship. Our political reality sums up the contradictory modernity of Latin America.

Our peoples chose democracy because it appeared to them to be the highroad to modernity. The truth is precisely the opposite: democracy is the result of modernity, not the path to it. The difficulties we have experienced in instituting democratic rule are one of the effects—the most serious one, perhaps—of our incomplete and defective modernization. But we were not wrong to choose this system of government: even with its many tremendous shortcomings, it is the best of all that humanity has invented. We have been mistaken, it is true, about the method for attaining it, since we have limited ourselves to imitating foreign models. The task awaiting Latin Americans, one requiring efforts of imagination at once bold and realistic, is to discover in our own traditions those seeds and roots—they are there—that will enable us to implant firmly and nourish a genuine democracy. It is an urgent task, and there is almost no time left. My warning is justified, for the traditional either/or of Latin America—either democracy or military dictatorship—is no longer relevant. In recent years a third term has appeared: the military-bureaucratic dictatorship that, through a colossal historical error, we call "socialism."

In order better to understand the terms of this historic either/or confronting our peoples, I can only repeat, briefly, some of the concepts of another essay, "Latin America and Democracy,"[6]

6. See in particular the section entitled "Historical Legitimacy and Totalitarian Atheology."

which appears in Part Two of this book. The political instability of our countries began on the morrow of our Independence. Unfortunately, historians have not explored the causes of this instability, or else have provided overhasty explanations. It is clear, in any event, that agitations and military coups in Latin America correspond to the violent upheavals and disturbances that have rocked Spain and Portugal since the nineteenth century. They are an integral part of a past that refuses to disappear: to modernize means to abolish this past. Although the United States did not create this instability, it took advantage of it from the very beginning, and also promoted it: without such instability the United States' domination might not have been possible. Another result of the hegemony of the United States was its withdrawal of us, so to speak, from universal history. During the period of Hispano-Portuguese domination, our countries lived on the periphery of the world, in an isolation that, as the historian Edmundo O'Gorman has pointed out, was fatal to our political education. Ever since then our peoples have been self-involved, like the Mexicans, or avid for novelties from outside, like the Argentines. The hegemony of the United States isolated us once again: the central problem of our State Departments lay in finding the best way of gaining Washington's friendship or of avoiding its interference. The curtain between Latin America and the world went by the name of the Monroe Doctrine.

Ever since Independence, despite military coups and dictatorships, democracy has been regarded as the only constitutional legality of Latin American nations—that is, as historical legitimacy. Dictatorships, as even the dictators themselves publicly conceded, were breaks in democratic legitimacy. Dictatorships represented the transitory, while democracy constituted the permanent reality, even if it was an ideal reality or one only imperfectly and partially embodied. The regime in Cuba very soon stood out as something quite different from traditional dictatorships. Although Castro is a leader within the purest tradition of Latin American *caudillismo*, he is also a communist leader. His regime lays claim to representing the new, revolutionary legitimacy. This legitimacy has replaced not only *de facto* military dictatorship but also the old historical legitimacy: representative democracy with its system of individual guarantees and human rights.

In order to make the implications of this novelty absolutely clear, I must underscore what I pointed out earlier: Latin American military dictatorships have never claimed to be replacements for democratic rule and have always been looked upon as transitory and exceptional governments. I am not attempting to absolve dictatorships: more than once I have condemned them. What I am trying to demonstrate here is their historical particularity. The Cuban regime lays claim to being a new legitimacy that permanently replaces democratic legitimacy. This novelty is no less important than the Russian presence in Cuba: it upsets the traditional perspectives of Latin American political thought and confronts this thought with realities that seemed inconceivable a generation ago.

As everyone knows, thanks to the takeover by the Castro regime in Cuba and to a series of chance circumstances—among which the key one was the arrogance and blindness of the United States government—Soviet power, without even having sought it, obtained as a gift from history a political and military base in America. Yet before Cuba's entry into the Soviet bloc, the independent policy of the revolutionary regime vis-à-vis Washington aroused the admiration and the well-nigh-unanimous fervor of the peoples of Latin America, as well as the friendship of many others. The Cuban revolutionaries— and this fact is very often forgotten—also succeeded in gaining the sympathy of a large part of U.S. public opinion, even though the United States government had long backed the mediocre and cruel dictatorship of Batista. But Washington—doubtless recalling its interventions in Guatemala, Santo Domingo, and Nicaragua—adopted a policy that was at once disdainful and hostile. Castro then sought Moscow's friendship. Of course, the faults and errors of various American administrations do not make Castro's victory over Batista a triumph of socialism. The classic Marxist thinkers had quite a different idea of what a socialist revolution ought to be.

Aside from the ideology underlying Castro's movement and the historical nature of his regime, the United States is today feeling the effects of its lack of understanding—more precisely, lack of sensitivity—in being confronted with the new and changed reality of Latin America. It has not only been obliged to accept the existence, not a great many miles off its shores, of a regime openly allied to Moscow, but has also been powerless to prevent Cuban troops, armed by the Soviet Union, from intervening in Africa, and Fidel Castro from undertaking frequent diplomatic offensives against the United States. Is the installation of the Castro regime in Cuba a first shadow of the twilight of U.S. hegemony? I couldn't really say whether it is a twilight or merely a dark passing cloud. What is certain, however, is that we are confronting an absolutely new situation in Latin America. For the first time in almost two centuries, a non-American power has a political and strategic base in this hemisphere. To realize the full historical import of the Russian presence in a Latin American nation, we need only recall its best-known antecedent: the abortive intervention of Napoleon III in Mexico in the middle of the last century.

The military dictatorships of Latin America have always resorted to a pretext in order to justify their existence: they are an exceptional, temporary remedy against the disorder and the excesses of demagogy or against threats from outside, or against "communism," by which was meant all nonconformists, dissidents, critics. But it is now plain that what communism—the real thing: the totalitarian system that has co-opted the name and the tradition of socialism—threatens first and foremost is democracy. The supposed historical justification of dictatorships is collapsing: by doing away with democratic rule, they pave the way for a totalitarian attack. I do not know if anyone has reflected upon the implications of the following: Fidel Castro overthrew not a democratic government but a corrupt dictator. In Central America today, it was not the tiny democracy of Costa Rica but a new regime in Nicaragua that replaced the dictator Somoza, a regime that is transforming itself day by day, before our very eyes, into a communist dictatorship. The only effective defense against totalitarianism is democratic legitimacy.

In "Latin America and Democracy" I mention the features that characterize the Central American situation: the fragmentation into tiny republics that are viable neither economically nor politically and that also lack a clear national identity (they are fragments of a body that has been torn to pieces); oligarchies and militarism, allied to U.S. imperialism and fostered by it; the absence of democratic traditions and the weakness of the middle class and the urban proletariat; the appearance of minorities of professional revolutionaries from the upper bourgeoisie and the middle class, many of them educated in the Catholic (most often Jesuit) schools of the bourgeoisie, radicalized by a series of circumstances that Freud would be able to explain though Marx would not; the intervention of Cuba and the Soviet Union, which have armed Nicaragua and trained guerrilla groups in El Salvador and other countries. . . .

At the time these lines are being written[7] it is impossible to predict what the outcome of the Central American conflict will be. Will the United States be able to resist the temptation to use force and to rely on military dictatorships and the extreme right? Will democratic groups, which can count on the support of the majority but are disorganized, be able to restructure themselves and win out over the extremists of left and right? Although such efforts almost always fail, perhaps a resolute course of action on the part of Venezuela, Mexico, and Co-

7. March 1983.

lombia, undertaken in conjunction with—why not?—socialist
Spain, could prevent a catastrophe and clear the way for a
peaceful and democratic solution.[8] Let us hope that it is not
too late. The establishment of communist dictatorships in Central America—a possibility that until now our governments,
with an insouciance that is beyond understanding, have badly
underestimated—would have terrible effects on Mexico's domestic peace and her foreign security. By their very nature,
these regimes are certain to be militias inspired by a bellicose
and expansionist ideology, and hence prepared to seek domination by violent means.

During the recent past, military dictatorships in South America
became more numerous, and those that already existed became stronger. At the end of this period, however, signs of a
return toward more democratic forms have appeared, a trend
particularly noticeable in Brazil. This is a phenomenon of immense import—providing, naturally, that the tendency continues and gathers strength. Brazil is destined to be a great power
and has already attained a considerable degree of development. Its conversion to democracy would contribute enor-
mously to changing the history of our continent and that of
the world. There are also other heartening symptoms. Vene-
zuelan democracy now gives every appearance of being a stable, healthy, and viable regime; like Costa Rica, Venezuela has
succeeded in creating a genuine democratic legitimacy. Deeply
shaken by the events of 1968, the regime in Mexico instituted
sensible reforms, and appreciable advances have been made.
The majority of Mexicans see in democracy, if not the remedy
for their tremendous problems, at least the best method for
discussing them and working out solutions. But since the case
of Mexico is unusual, it is not beside the point to dwell for a
moment on the situation in that country.

On July 4, 1982, Mexicans elected a president and new sen-

---

8. This was written before the so-called Contadora group, which also includes
Panama, was formed.

ators and deputies. The elections were notable both for the large number of voters and for the freedom and orderliness with which the voting took place. The Mexican people showed once again that their political morale is better and healthier than that of the classes that lead them: the bourgeoisie, the professional politicians, and the intellectuals. Two months later, on September 1, those same classes again gave proof of the weakness of their democratic inclinations. The country was— and still is—confronted with a disastrous economic situation. The causes are well known: the deterioration of the worldwide economy (inflation, unemployment, the fall of the prices of petroleum and commodities, high interest rates, etc.); the rash and improvident policies of the Mexican government, which once again turned a deaf ear to those of us who had repeatedly expressed our concern about the careless way in which the tremendous wealth from the newly discovered oil deposits was being handled;[9] and the endemic disease of patrimonialist regimes such as the Mexican: the corruption and venality of government officials. The flight of capital—a consequence, not a cause, of the country's ills—precipitated the financial crash. Although the outgoing president had only three months left in office, his answer to the crisis came like a thunderbolt: nationalize banking; or, rather, place it under government control, since it was already Mexican. The measure was adopted without consultation and without prior discussion. Since it was decided on in secret, it surprised everyone when it was made public on September 1, including the majority of the ministers of state, among them the minister of finance himself.

Having barely recovered from the shock, popular opinion was given no chance to express its opposition. The government and its political organ, the PRI,[10] marshaled all of its immense propaganda resources in support of the measure. The

9. See the prologue of *El ogro filantrópico* (1979).
10. The Partido Revolucionario Institucional, the Institutional Revolutionary Party (formerly the National Revolutionary Party), which has governed Mexico since 1929. (TRANS.)

communications media, some out of conviction and others out of fear of being placed under government control in their turn, joined the official chorus or maintained a discreet silence. At the same time, the leaders of the labor-union bureaucracy also mobilized the workers. But the real indication of the state of public morale was the reaction of independent groups: bankers and management protested timidly; the parties of the left and their intellectuals applauded enthusiastically. Only a few dared to criticize the president's decision: a few journalists, three or four intellectuals, and AN,[11] the opposition party.[12] Different views, favorable and unfavorable, can be taken about the state takeover of banking; what was reprehensible was the way in which it was decreed, a mixture of the surprise attack at dawn and a drumhead court verdict. The silence of some and the paeans of praise of others were more disgraceful still.

Why did bankers and management complain *sotto voce* and thus fail to be heard by popular opinion? For one thing, because many of them had been faithful props of the regime; more important, none of them had ever bothered to try to improve the political culture of Mexico or ever lent a hand in the task that is incumbent not just on one class but on all Mexicans: creating a political space that is truly free and open to all tendencies. They were and are pressure groups, not sectors of opinion. How could they have called upon democratic principles to aid and protect them if they had never lifted a finger to defend those principles and implant them deeply in our public life?

The attitude of the left and of its intellectuals was no less deplorable, especially to those who remembered their recent

11. Acción Nacional, National Action. (Trans.)
12. The most complete and penetrating analysis of this event is to be found in two essays published in the review *Vuelta*, one by Enrique Krause, "El timón y la tormenta" ["The Helm and the Storm"] (no. 71, October 1982), and the other by Gabriel Zaid, "Un president apostador" ["A President Who Gambles"] (no. 73, December 1982).

noisy professions of pluralist and democratic faith. Like the Hebrews fascinated by the golden calf, they returned to their idolatry of the state following the presidential decree. Instead of censuring the way in which banking had been placed under state control, they hastened to hail the measure, as though it were a truly revolutionary act. They did not ask themselves the basic question that all of their teachers would have asked: what social group does it benefit? It is clear that those who gained from it were not the workers or the people in general but the new class—that is, the state bureaucracy. The power of the government, which in Mexico is already excessive, was thereby strengthened.

The elections of July 1982 demonstrated that the majority of Mexicans are inclined toward democratic solutions; the events of September confirmed that neither the conservative bourgeoisie, nor the parties of the left and its intellectuals, nor the governmental political class has a genuine democratic vocation. These groups are captive, some of them prisoners of their interests and others of their ideology. In order to understand the lack of independence on the part of capitalist managers and the leaders of the labor unions, my readers must remember that both have prospered in the shadow of the Mexican State, which has been the agent of the modernization of the country. (I have dealt with this theme in *El ogro filantrópico*.) [13] Until recently, the intelligentsia has been part of the bureaucracy; only in the last few years have intellectuals found posts opening up for them in the universities, which have increased both in numbers and in size. The function once filled by the Church and the religious orders is now being filled by the universities, a parallel that is even more striking if we note that the universities are public institutions closely associated with the State, as was the Church in New Spain. It can even be said

13. The title essay of this volume appeared in English as "The Philanthropic Ogre," *Dissent*, Winter 1979.

that their dependence is greater: the Church was immensely wealthy until the middle of the nineteenth century, whereas Mexican universities survive on government grants and subsidies. The relationship between Mexican intellectuals and the State is no less ambiguous than that between the clerics of New Spain and temporal power: it would not be inexact to define it as *conditional independence*.

The lack of communication between the real country and the classes that lead it, not excluding intellectuals, is a characteristic and persistent phenomenon in the modern history of Mexico. The people have not succeeded in giving expression to their grievances and their needs in a coherent line of political thought or in realistic programs because the sort of intellectual and political minorities that in other countries interpret and give concrete form to vague popular aspirations are in our country hypnotized by simplistic ideologies. Those intellectuals who are not catechumens of the churches and sects of the left defend the *status quo*, where their interests lie. Others, a minority, pursue their own specific activities—research, teaching, artistic and literary creation—and thus preserve the fragile continuity of our culture, today more threatened than ever. But the independent intellectuals who have taken on the function of criticizing and who dare to think for themselves are few and far between—a mere handful, really. Another obstacle is that the communications media are controlled, directly or indirectly, by the government, and the influence of ideologues in the daily and weekly press is preponderant. Hence rumor and jokes are the most common forms of popular expression. To make myself clear: there is not a governmental dictatorship over opinion, but there is a lack of communication between the real Mexico and those who should be its spokesmen and interpreters. Despite all these adverse circumstances, however, public opinion is coming more and more to reject the patrimonialism and paternalism of the regime, and aspires to a free and democratic public life. It would be very dangerous if our leaders were to disregard this insistent, general outcry.

The same evolution can be noted on the rest of the continent. In Colombia, democracy is not just holding its ground: by persisting, it is advancing. Peru and Ecuador are returning to democratic forms, and the same thing is happening in Bolivia. In Uruguay, elections took place whose results were unfavorable to the military. Those Chileans who went into exile are beginning to come back home, a sign that perhaps a change in that country is not too far off. In Argentina, there are indications of a return to democracy in the offing. In short, it does not seem too bold to conclude that we have arrived at a turning point in history. If the tendency that I have very briefly described grows stronger and spreads, we Latin Americans could begin to think of joint democratic actions that would answer the real interests of our peoples. Until now we have gone along with the game of the Great Powers. The time has come to try a continental policy that would be new and that would be our own. Perhaps it is not at all illusory to think that a considerable contribution to this renaissance might be made by two new European democracies to which our history and our culture unite us: Spain and Portugal. I believe that opportunities are opening up for a continental democratic action, coming not from outside but from our countries themselves. An alliance of democratic Latin American nations would not only make Washington stop and think but could actually change our continent. I am thinking of two essential contributions it could make: furthering the real modernization of our peoples, the institution of true democracies; and re-establishing Latin American independence, this time on firmer foundations. In the modern world, democracy and independence are closely linked: a democracy that is not independent is not a true democracy. But we do not have much time: darkness is falling, and the skies are still threatening.

An Ink Blot

Beneath a gentle sun, the quiet hours passed one by one. The day was ending, and, in a moment of indecision and distraction, I was idly following a thread of light that fell on the papers on my desk. Through the window a blur of azaleas, motionless in the late-afternoon light. Suddenly I saw a shadow rise from the written page, move toward the lamp, creep over the reddish cover of the dictionary. The shadow loomed larger and turned into a figure that might have been human or titanic: I could not say which. Nor could I tell what size it was; it was at once minuscule and immense. It was making its way about among my books, and its shadow extended over the whole universe. It looked at me and spoke. Or, rather, I heard what its eyes said to me—though I do not know if it had eyes and if those eyes were looking at me:

HE: So you've finished *Threatening Skies*?
I (*nodding my head*): Who or what are you?
HE: My name is Legion. Endlessly changing name and form, I am many and yet no one; I am forever imprisoned within myself and yet I cannot grasp myself. A Byzantine called me Lucifugus: an errant darkness, enemy of the light. But my shadow is light, like that of Aciel, the black sun. I am light streaming inward, light in reverse. Call me Eçul.[14]
I: I know who you are and where you come from.
HE: Yes, I come from the First Book of that work (*and he points to a volume with a half binding*). But I am not mentioned by name. I am one of the retinue.
I: Of whom?
HE: Of a prince. His name would mean nothing to you.
I: Why have you come?

14. Luce = light.

HE: To dissuade you. You have been wandering about, lost in time—or, as you people say, in history. You are seeking your way; have you found it?

I: No. But I now know that revolts petrify into revolutions or are transfigured into resurrections.

HE: That's nothing new. It's as old as you people's presence on this planet.

I: It's new to me. New to us, the ones who are alive now.

HE: Does the contention of the two powers strike you as being new, too? Remember Rome and Carthage. . . .

I: No, that's nothing new. And yet it isn't the same. Similarity is not identity.

HE: What an illusion! Haven't you noted another similarity that's even more . . . impressive?

I: Which one is that?

HE: Do you remember the assembly in Pandemonium, here in this book? *(He points to the volume once again.)*

I: I don't understand what you mean.

HE *(impatiently)*: The earth is now marked off into hopscotch squares [15] in which we devils play at warring with men.

I: I tell you I don't understand.

HE: You haven't read your poets carefully. *(Didactically)* In Pandemonium there are two great princes, inferior in power only to Lucifer. The first is a *(he begins to recite)* "horrid King besmear'd with blood of human sacrifice, and parents' tears, Though for the noise of Drums and Timbrels loud, Their children's cries unheard, that pass'd through fire To his grim idol."

I: Ah, Moloch, the god of the Ammonites!

HE: And of the Israelites. Don't you know *(he lowers his voice)* that the deity to whom they sacrificed children was Yahweh? This happened in the days of Achaz and Mannasseh. Your Biblical studies are extremely sketchy. You might know him by another of his names, however: Aries.

15. One of the Spanish names for hopscotch is *infernáculo*. (TRANS.)

I: Mars.

HE: Huitzilopochtli, Tezcatlipoca, Odin, Thor . . .

I: And Kali, who licks battlefields clean with her immense tongue and rakes cemeteries with her fingernails.

HE: The other prince doesn't go about with head held high like the great Moloch; his is always bowed, looking downward. Humility? No: he is searching, looking for hidden treasures. He taught you people to explore—the poet says "plunder" but he exaggerates—the bowels of the earth. When we lived on high (*he begins reciting once again*), "his hand was known In Heav'n by many a Tow'red structure high Where Scepter'd Angels held their residence . . ." until, cast down from the crystalline battlements, "from Morn to Noon he fell, from Noon to dewy Eve, A Summer's day . . . nor did he scape By all his Engines, but was headlong sent With his industrious crew to build in hell." (*He pauses.*) Since then he's limped a bit. . . .

I: Hephaestus, Vulcan . . .

HE: Mammon is his real name. Patron of smiths, tradesmen, engineers, mechanics, bankers, miners . . . An astute, enterprising, hard-working god. A demanding god. Matthew said: "Ye cannot serve God and Mammon."

I: A devil versed in the Scriptures!

HE (*disregarding my remark*): In Pandemonium there was once an argument—and anything that is once argued about there is argued about forevermore, because succession and repetition, change and identity are one and the same thing to us—over how we could recover the good that had been lost. It was the day after the Fall. "The strongest and the fiercest spirit that fought in heaven," Moloch, rose to his feet and said: "How, fugitives from heaven, can we remain seated here and accept as our abode this dark opprobrious den of shame when armed millions await the signal to assault the heavens? The tyrant reigns there above only through our delay. . . ." Moloch urged insurrection and war. Then Belial spoke, counseling prudence. But the discourse that surprised us all, and surprises us still,

was Mammon's—although the brow of the Evil One furrows
every time I mention it.

I: Why?

HE: Mammon proposed neither rebellion, as did the fiercely
proud Moloch, nor submission, as did the hypocritical Belial:
"Since we cannot overthrow the Almighty or obtain his par-
don (and even if this latter were possible: who would not feel
humiliated to spend his days celebrating Him with forced hal-
lelujahs? What a wearisome Eternity would be ours if we spent
it worshiping what we hate . . .) . No, let us not strive to do
the impossible nor resign ourselves to our unacceptable state
of splendid vassalage: we must rather seek our own good from
ourselves and live free in this vast recess, to none accountable,
preferring hard liberty to the easy yoke of servile pomp. . . .
Do we dread the deep world of darkness? We will traverse it
with lights imitating his. Does this desolation dishearten us?
We have the skill and art to raise magnificence. . . . Our very
tortures may, in time, turn to a second nature and the fires
that torment us be caresses. . . ." Mammon's harangue raised
a thunder of applause in Avernus like unto that of the storm
when it shakes the sea and makes the rocks and hollows of
the cliffs resound. Then Beelzebub . . .

I: I understand now why Satan was perturbed. Mammon's
speech was deviationist. With his shrewd program of reforms
he was doing his best to divert the infernal populace from their
most urgent task and, if I may so put it, from their mission in
history: insurrection and the seizure of heaven. . . . What I
still fail to understand is what the similarity is.

HE: You're thick-skulled. Think of Moloch and his rallying cry:
Let us all unite to assault heaven with millions and millions of
fanatical rebels. What does all that remind you of?

I: Russia, of course. It's a vassal of Moloch's!

HE: Correct. Now think of Mammon and his idea of making
hell habitable through work, industry, commerce, and hard
liberty.

I: The United States and the capitalist democracies! They're colonies of Mammon's!

HE: The history of humans is the acting out . . .

I: . . . of the devil's disputation.

HE: *Ecco!* You've gotten the lesson through your head.

I: Wait a minute, though. Even if it were true that history is only a play written by you devils, it would still be necessary to choose the lesser evil.

HE (*shocked*): Evil is neither greater nor lesser. Evil is evil.

I: You're telling me that, you who invented negation and divided eternity with it and made it successive time, history? You're talking like Gabriel and Michael, warriors of the Absolute.

HE (*in a conciliatory tone of voice*): We are condemned to live in time. We are eternal and we keep falling forever, yes forever, into the relative: that is our punishment. But not you people. In one leap you can escape time and its demoniacal quarrels. Don't you call that freedom?

I: What do you mean? Moloch proclaims his battle an absolute one, Mammon decrees that wealth is the supreme good; you devils have always made an absolute of the relative, a God of the creature, a false eternity of the instant. And now you tell me the contrary: the relative is relative forevermore and is demoniacal. You ask me to give up terrestrial disputes and look upward. Another trick, another trap.

HE: You are still imprisoned in time. Remember: "Nothing undeceives me/ the world holds me spellbound."[16] It is necessary to let go, take the leap, be free. The word is "detachment."

I: Trickster! We live in time and must confront time squarely. Only in that way will we perhaps glimpse nontime one day. Politics and contemplation: that was what Plato said and what was said after him, each in his own way, by Aristotle and Marcus Aurelius, Saint Thomas, Kant. In the relative there are traces,

16. Francisco de Quevedo.

reflections of the absolute; in time each minute is a seed of eternity. And even if this were not so, it doesn't matter: each relative act points to a meaning that transcends it.

HE: Dabbler in philosophy!

I: To each his own . . . We live in time, we are made of time, and our works are time: they pass away, as do we. But at times we can see a clearing in the cloudy sky. Perhaps there is nothing behind it, and what it shows us is its own transparency.

HE (*eagerly*): And is that enough? Is that reflection of a reflection enough for you?

I: It's enough for me, it's enough for us. We're the opposite of you: we can't renounce either action or contemplation.

HE (*in another tone of voice altogether*): For us, seeing and doing are the same thing—and lead nowhere. All our eloquent discourses end in serpent hisses. . . . We are spirits fallen in time, but we are not time: we are immortal. That is our doom—eternity without hope.

I: We are children of time and time is hope.

On the other side of the window, the azaleas had melted into the night. On the sheet of paper, in a blank space between two paragraphs, I noticed a tiny ink blot. I thought: a black hole.

Part Two

As Time Goes By

I

The Telltale Mirror

Before becoming a reality, the United States was an image to me. That is not surprising: we Mexicans begin as children to see that country as the *other*, an *other* that is inseparable from us and that at the same time is radically and essentially alien. In the north of Mexico the expression "the other side" is used to mean the United States. The other side is geographical: the frontier; cultural: another civilization; linguistic: another language; historical: another time (the United States is running after the future while we are still tied to our past); metaphorical: it is the image of everything that we are not. It is foreignness itself. Yet we are condemned to live with this foreignness: the other side is right next to us. The United States is always present among us, even when it ignores us or turns its back on us: its shadow covers the entire continent. It is the shadow of a giant. To us this giant is the same one that appears in fairy tales and legends—a big overgrown fellow and something of a simpleton, an ingenuous sort who doesn't know his own strength and who can be fooled, though his wrath can destroy us. The image of the goodhearted, doltish giant is juxtaposed with that of the shrewd and bloodthirsty cyclops. A childish image and a licentious one: the ogre that eats children up alive in Perrault, and the ogre of Sade, Minsk, at whose orgies the libertines eat smoking platefuls of human flesh, using the seared corpses as tables and chairs. Saint Christopher and Polyphemus. And Prometheus as well—the fire of industry and of war. The two faces of progress: the automobile and the bomb.

The United States is the negation of what we were in the sixteenth, seventeenth, and eighteenth centuries, and of what, since the nineteenth century, many among us would prefer us to be. Like all countries, Mexico has been born several times: the first time in the seventeenth century, a hundred years after the Conquest; the second at the beginning of the nineteenth century: Independence. But perhaps it is not correct to say that Mexico has been born twice; what we are really doing is calling several quite different historical entities by the same name (Mexico). The first of these entities is the old, indigenous society composed of city-states, ruled by military theocracies, that were the creators of complex religions and no less complex artistic works. Rather than another society, this world is seen as another civilization. Later, after the great cleft of the Conquest and Evangelization, around the middle of the seventeenth century, another society appears: New Spain. This society was not really a colony, in the strict sense of the word, but a kingdom subject to the crown of Spain like the others making up the Spanish Empire: Castile, Aragon, Navarre, Sicily, Andalusia, Asturias.

From the very beginning, the offspring of the Spaniards born in Mexico, the so-called *criollos*, felt different from Europeans, and their feeling of difference grew stronger from the seventeenth century on. Their awareness—initially dim and confused—of their own social and historical uniqueness found expression, slowly but powerfully, in the course of the seventeenth and eighteenth centuries. The originality of the Creole genius was particularly notable in three domains: sensibility, aesthetics, and religion. As early as the end of the seventeenth century, there was what we may speak of as a Creole *charac-*

ter—a manner of being in which a whole series of vital attitudes have developed, ways of confronting this world and the other, sex and death, leisure and work, oneself and others (above all those who were *others* by antonomasia: the European Spaniards). In the realm of the arts, the Creoles expressed themselves with great felicity; I need hardly mention Baroque architecture—both in its erudite and in its popular form—or the figure of Sor Juana Inés de La Cruz. This woman was not only a great poet but also the intellectual conscience of her society and, in certain respects, of our own society today. But the great creations of New Spain were above all in the sphere of religious beliefs and myths. And the greatest of all of these was the Virgin of Guadalupe.

Though Creole society had had separatist aspirations ever since its birth, these did not manifest themselves openly until the last years of the eighteenth century. The embryonic, confused nature of the Creoles' political sentiments stands in sharp contrast to the richness, complexity, and originality of their artistic and religious creations and expressions. When the Creoles began to think in political terms, they were inspired by the Jesuits. The Society of Jesus had made itself not only the educator of the leading Creole class but also its moral and political conscience. At the very time when the revelation of their uniqueness was coming to political consciousness, certain Creoles realized that their tradition—monarchism and Neo-Thomism—had not only lost almost all its vitality but also did not offer a broad enough base to serve as the solid underpinning of their aspirations and enable them to articulate these in a program of political action. Of course, the Creoles and their Jesuit mentors did have in mind a vague project for an empire of North America. This idea, whose origins go back to the last third of the seventeenth century, lingered on in the Mexican Conservative Party until the middle of the nineteenth.

More than a political idea, the Mexican Empire was an image. As an image, it attracted many outstanding minds; as an idea, it revealed certain inconsistencies. The first and gravest of these was that, since it was a simple extension of the Spanish system, it did not offer the basic elements and concepts needed to draw up a truly national program. In truth, New Spain was a social reality much vaster than the Creoles, comprising social and ethnic groups, Indians and mestizos, whose hearts could not be stirred by the idea of empire. Apart from being a mere extension of the Spanish system, the idea of a Mexican Empire was a project that ran counter to the general current of that era. In those years the idea of a nation, which was the foundation of independence, had become inseparable from the idea of popular sovereignty, which was, in turn, the foundation of the new republics. Finally, there was in the aspiration to independence an element that did not appear in the imperial project: the eagerness for modernity. Or, as people put it in those days, the eagerness for progress. In Mexico and in the rest of Hispanic America, "independence," "republic," and "democracy" were words synonymous with "progress" and "modernity."

The expulsion of the Jesuits precipitated the intellectual crisis of the Creoles: not only were they left without teachers but also without a philosophical system to justify their existence. Many of them then turned to the other tradition, the enemy of the tradition that had founded New Spain. At that moment the radical difference between the two Americas became visible and palpable. One, the English-speaking one, is heir to the tradition that has founded the modern world: the Reformation, with its social and political consequences, democracy and capitalism; the other, ours, the Portuguese- and Spanish-speaking one, is heir to the universal Catholic monarchy and the Counter-Reformation. The Mexican Creoles could not found their separatist project on their own political and religious tradition: they *adopted*, though without *adapting* them, the ideas of the other tradition. This was the moment of the second birth of Mexico; more precisely, it was the moment when New Spain, in order to consummate its separation from Spain, denied itself. That negation was its death and the birth of another society—Mexico.

The United States enters our history during this second moment, making its appearance not as a foreign power that must be fought, but as a model that must be imitated. This was the beginning of a fascination that, despite having changed form during the last 150 years, is still as intense as ever. The history of that fascination goes hand in hand with the history of the groups of intellectuals who, since Independence, have drawn up all those programs of social and political reform aimed at transforming the country into a modern nation. Above and beyond their differences, there is a shared idea that inspires liberals, positivists, and socialists alike: the project of modernizing Mexico. Since early in the nineteenth century this project has always been defined vis-à-vis—whether for or against—the United States. The passion of our intellectuals for U.S. civilization ranges from love to bitter rancor, from adoration to horror. These are contradictory but coincident forms of ignorance: at one extreme, the liberal Lorenzo de Zavala, who did not hesitate to side with the Texans in their war against Mexico; at the other, the contemporary Marxist-Leninists and their allies, the "liberation theologians," who have made of materialist dialectic a hypostasis of the Holy Spirit, and of American imperialism the prefiguration of the Antichrist.

The "intelligentsia" is not the only class to have experienced contrary feelings toward the United States. On the morrow of Independence and up until the middle of the nineteenth century, the well-to-do classes were resolutely against it; later on they became its allies and, in almost every case, its servants and accomplices. Nonetheless, since in the final analysis they are heirs of the hierarchical society that was New Spain, our rich have never really wholeheartedly adopted the liberal and democratic ideology; they are friends of the United States for reasons of interest, but their real moral and intellectual affini-

ties lie with authoritarian regimes. Hence their sympathy for
Germany during the two world wars of this century. The same
evolution can be seen in the political and military castes: General Miramón, a conservative, was an enemy of the United
States, but General Porfirio Díaz, a liberal, was his proconsul.
It may be concluded, insofar as it is possible to venture gen-
eralizations in so contradictory a domain, that during the nineteenth century the liberals were the friends and allies of the
United States (the outstanding example is Juárez) and the conservatives its adversaries (Lucas Alamán is the most notable
case), whereas in the twentieth century these roles have been
reversed. But in both centuries the enemies of the Americans
have had to seek allies and protectors outside the continent:
in the nineteenth century, a Miramón looked toward France;
in the twentieth, a Fidel Castro looks toward the Soviet Union.

Brazil did not repudiate Portugal; in Hispanic America, on
the other hand, the liberals were anti-Spanish. The anti-His-
panolism of our liberals may appear to be absurd and irra-
tional, and in fact it is. But it is also explainable: the democratic
ideas adopted by the liberals were the negation of everything
that New Spain had been. The Revolution of Independence, in
Mexico and in all of Spanish America, was simultaneously an
affirmation of the Hispano-American nations and a negation
of the tradition that had founded these nations—a self-nega-
tion. Here there appears another difference from the United
States. On declaring their independence from England, the
Americans did not break with their past; on the contrary, they
affirmed what they had been and what they wanted to be. The
Independence of Mexico was the negation of what we had
been since the sixteenth century; it was not the institution of
a national project but the adoption of a universal ideology foreign to the whole of our past.

Between puritanism, democracy, and capitalism there was
not opposition but affinity; the past and the future of the United
States are reflected without contradiction in these three words.
Between republican ideology and the Catholic world of the

Mexican viceroyalty, a mosaic of pre-Columbian survivals and Baroque forms, there was a sharp break: Mexico denied its past. Like all negations, ours contained an affirmation—that of a future—except that we did not work out our idea of the future using elements and concepts drawn from our own tradition; instead we took over ready-made the image of the future invented by Europeans and Americans. Ever since the seventeenth century, our history, a fragment of Spain's, has been an impassioned negation of the modernity being born: the Reformation, the Enlightenment, and all the rest. As the twentieth century dawned, we decided that we would be what the United States already was: a modern nation. The entry into the Modern Age demanded a sacrifice—of ourselves—and the result of this sacrifice is well known: we are still not modern, but have been searching for ourselves ever since.

The first seeds of democracy on this continent appeared among the dissident communities and sects of New England. It is true that the Spanish established in the countries of the Conquest the institution of the *ayuntamiento*, the municipal council, founded on the principle of self-government of cities and towns. But the existence of the *ayuntamientos* was always precarious, strangled as they were by a vast, complex network of bureaucratic, nobiliary, ecclesiastic, and economic jurisdictions and privileges. New Spain was always a hierarchical society, without representative government and dominated by the dual power of the viceroy and the archbishop.

Max Weber divided premodern regimes into two great categories: the feudal system and the patrimonial. In the first, the prince governs with—or at times against—his equals by birth and rank: the barons. In the second, the prince rules the nation as though it were his patrimony and his household; his ministers are his family and his servants. The Spanish monarchy is an example of a patrimonialist regime. So were (and are) its successors, the "democratic republics" of Latin America, forever wavering between the Caudillo and the Demagogue, the despotic Father and the rebellious Sons.

From their very beginning, the religious communities of New England jealously declared their autonomy with respect to the State. Inspired by the example of the Christian churches of the first centuries, these groups were always hostile to the authoritarian and bureaucratic tradition of the Catholic Church. Since Constantine, Christianity had lived in symbiosis with political power; for more than a thousand years the Church had followed the imperial-bureaucratic model of Rome and Byzantium. The Reformation marked the rupture of this tradition, a rupture that the religious communities of New England, in their turn, carried to its limits, emphasizing the egalitarian features and the tendency toward self-government of the Protestant groups of the Low Countries.

In New Spain the Church was first and foremost a hierarchy and an administration—that is, a bureaucracy of clerics that in certain respects is reminiscent of the institution of the mandarins of the old Chinese Empire. Hence the admiration of the Jesuits when, in the eighteenth century, they were confronted with the regime of K'ang-hsi, in which they at last saw the concrete realization of their idea of what a hierarchical and harmonious society could be—a society that was stable but not static, like a watch that keeps perfect time and doesn't run down. In the English colonies the Church was not a hierarchy of clerics who were the exclusive repositories of knowledge, but the free community of the faithful. The Church was plural and from the beginning it was made up of a network of associations of believers, a true prefiguration of the political society of democracy.

The religious basis of American democracy is not visible today, but it is not any less powerful in consequence. More than a foundation, it is a buried taproot; the day it dries up, so will that country. If we do not take this religious element into account, it is impossible to understand either the history of the United States or the meaning of the crisis that it is undergoing today. The presence of the Protestant religious ethic transforms an incident such as Watergate into a conflict that shakes the very foundations of American democracy. These foundations are not only political—the social contract between men—but religious: the covenant of men with God. In all societies politics and morality exist side by side, but unlike what happens in a secular democracy such as that of France, in the United States it is almost impossible to separate morality and religion. In France democracy was born of the criticism of the two institutions that the *ancien régime* represented: the Throne and the Altar. The consequence of the criticism of religion was the rigorous separation between religious morality, a private domain, and political morality. In the United States, on the other hand, democracy is the direct offspring of the Reformation— that is, of a *religious* critique of religion.

The fusion between morality and religion is characteristic of the Protestant tradition. In reformed sects, rites and sacraments yield their cardinal place to morality and soul-searching. Other eras and other civilizations had known theocracies of warrior-monks and empires ruled by priestly bureaucracies; the union of theology and power, dogma and authority, was a frequent phenomenon, not a rarity. It was the Modern Age, the age that made the critique of the kingdom of heaven and its ministers on earth, that undertook to invert the terms of the age-old, impure alliance between religion and politics. American democracy lacks dogma and theology, but its foundations are no less religious than the covenant that unites Jews and Jehovah.

Because of its religious origins and also because of the political philosophies that later shaped it, American democracy tends to strengthen society and the individual vis-à-vis the State. From the very beginning, we find in U.S. history a twofold aspiration, toward egalitarianism and toward individualism. Seeds of life, but at the same time seeds that are contradictory. In recent days, American intellectuals, on the occasion of the Bicentennial of Independence and in the face of the crisis that has shaken

the very pillars of their nation, have again posed the question
that divided the Founding Fathers: freedom or equality? The
controversy risks becoming a Scholastic disputation: from the
moment they lose their concrete historical dimensions, freedom and equality become entelechies. Freedom is defined in
terms of its limits and the obstacles that confront it; the same
thing happens with equality. In the case of the United States,
freedom was defined in contradistinction to the hierarchical
inequality of European society, so that its content was egalitarian; the desire for equality, in turn, was manifested as con-
certed action against the oppression of economic privileges—
that is, as self-determination and freedom. Both freedom and
equality were subversive values, but that was because they had
previously been religious values. Freedom and equality were
dimensions of the life beyond: they were gifts of God, appear-
ing mysteriously as expressions of divine will. Just as in Greek
tragedy the freedom of heroes is a dimension of Destiny, so in
Calvinist theology freedom is closely linked to predestination.
Hence the religious revolution of the Reformation anticipated
the political revolution of democracy.

In Latin America, exactly the contrary occurred: the State
fought the Church, not to strengthen individuals but to replace
the clergy as the power controlling consciences and wills. In
our America there was no religious revolution to pave the way
for political revolution; nor was there, as in eighteenth-
century France, a philosophical movement to take on the criticism of religion and the Church. The political revolution in
Latin America—I am referring to Independence and the struggles between liberals and conservatives that drenched our
nineteenth century in blood—was merely a manifestation, yet
another, of Hispano-Arabic patrimonialism: it did battle with
the Church as with a rival whose place must be taken over; it
fortified the authoritarian State, and the liberal *caudillos* were
no more benign than the conservatives; it accentuated centralism, though beneath the cloak of federalism; and, finally, it

made an endemic disease of the exceptional regime that has ruled in our lands since Independence: *caudillismo.*[1]

Independence was a false beginning: it freed us from Madrid, but not from our past. To the evils we had inherited we added others that were all our own. As our dreams of modernization faded, the fascination exerted on us by the United States grew. The war of aggression of 1847 turned it into an obsession. An ambivalent fascination: at one and the same time, the titan was the enemy of our identity and the unacknowledged model for what we would most like to be. In addition to being a political and social ideal, the United States was an interventionist power, an aggressor. This double image, which corresponded then, and still corresponds today, to a reality—the United States is both a democracy and an empire—lingered throughout the nineteenth century and was one of the basic themes of the bitter polemic between liberals and conservatives. In order to understand the attitude of the liberals, it suffices to recall what the attitude of thousands of "progressivist intellectuals" has been toward the Soviet Union: they close their eyes to the reality of the vast Soviet bureaucracy, its omnipresent and omnipotent police, its concentration camps, and the imperialist policy of Moscow, seeing only their mental image of a free, peaceful, beatific socialist fatherland. Though liberals were less credulous, in their case, too, ideology was more real than reality itself.

The liberals were enemies of the Mexican past, which they denounced as a foreign imposition, a Spanish intrusion. Thus they said that Mexico had "regained" its Independence, as though the nation had existed before the arrival of the Spaniards. What they celebrated by contrast to the inauthentic past of the viceroyalty was not the pre-Columbian past—which they

---

1. Rule by *caudillos*, who were military leaders, often of irregular forces, and, by extension, strongmen, political bosses. (TRANS.)

had almost no knowledge of, and which many of them scorned: Juárez, an Indian, was not an *indigenista*[2]—but, rather, the future of liberal democracy. In the face of the two eccentricities (from the modern Western point of view) that had made us what we were, the Spanish past and the Indian past, the liberals postulated an abstract universality, derived from the progressivist ideologies of the day. The United States was the immediate example of this universality: in its present we could see a vision of our future. A telltale mirror: like the stepmother's mirror in the fairy tale, each time we asked it to show us our image, it showed us that of the *other*.

Conservatives thought that the United States, far from being a model, was a threat to our sovereignty and our identity. The ideological contagion seemed no less dangerous to them than the physical aggression: it was a complementary form of penetration. If the future that the liberals proposed to us was an alienation, the defense of our own present demanded, for the same complementary reason, that of our past. The conservatives' reasoning was only apparently irreproachable: the past that they were so stubbornly defending, and, not without reason, identifying with the present, was a past in the process of disappearing. I have already pointed out that the tradition of New Spain—an admirable tradition, which modern Mexicans have had the stupidity to ignore and even to scorn—offered no elements or principles that could serve to resolve the twofold problem confronting the nation: that of independent life and that of modernization. The first had to do with finding the political form and the social organization that an independent Mexico should adopt; the second, with working out a viable program that, without too many upheavals, would permit the country to enter upon that modernity to which, until then, the Spanish Empire had barred the way. Like all of Spanish Amer-

---

2. In this context, a defender of native Indian peoples and their rights. The term is also used for writers and scholars dealing with Indian themes and ways of life.

ica, Mexico was condemned to be free and to be modern, but its tradition had always rejected freedom and modernity.

Independence was not so much a consequence of the triumph of liberal ideas—adopted only by a minority—as of two circumstances that to date have received very little attention. The first was the disintegration of the Spanish Empire. It must here be repeated once again, even though the statement may cause more than one reader to raise an eyebrow, that the Independence of Spanish America was achieved not only because of the action of the insurgents but also, and most significantly, because of the inertia and paralysis of the mother country. The continent of Hispanic America was young and rebellious, but healthy: Spain was a soul fast asleep in a body drained of its lifeblood. This image accounts for the phenomenon of Independence with greater economy and no less exactitude than ideological explanations. The second crucial circumstance was the existence of a social contradiction in New Spain. This contradiction, which was insoluble within the presuppositions underlying the social order of New Spain (and was not without certain analogies to that faced by the United States today with respect to its ethnic minorities), lay in the opposition between Creoles and mestizos. The leading class of Creoles postulated an abstract universalism, but the mestizos—the new historical reality—had no place outside of a metaphorical one in this universalism. The heirs to the twofold Hispanic universalism (the Empire and the Church), the Creoles, under the influence of the Jesuits, had dreamed in the eighteenth century of a Mexican Empire: New Spain would be the Other Spain. Independence fulfilled the separatist ambitions of the Creoles, except that it was not they who were the real winners, but, rather, the mestizos, who thus far had been a marginal group within the society of New Spain. Mexico was not an empire but a republic, and the ideology that nourished its governing caste was not that of a Catholic empire but bourgeois nationalism.

The episode of Maximilian is a cruel illustration of the illu-

sory nature of the conservatives' project. Calling upon a European prince to found a Latin empire that would check the
expansion of the Yankee republic was a solution that was not
at all preposterous in 1820 but had become an anachronism
by 1860. The monarchical solution had ceased to be viable,
because the monarchy was identified with the situation prior
to Independence. The perceived difference between Creoles and
Spaniards had been the decisive cause of the separatist movement. From the very outbreak of hostilities, mestizos and many
Indians—that is to say, the dispossessed classes—had participated in the battles for Independence. It was only natural that
once Independence had been attained, these groups should have
no interest in reconstituting the monarchical system—not out
of republicanism, but because monarchy represented the legi-
timization and consecration of the existing social hierarchies.
The mestizos, who were the most energetic and dynamic sector of society, were seeking their place in the sun, socially,
economically, and politically. The democratic republic opened
doors to their aspirations and ambitions, even though these
latter had little to do with either republicanism or democracy.

The liberal ideology was not a real solution. The nationalism
of the republicans was a superficial imitation of French nationalism; its federalism—a copy of the American—was a disguised *caciquismo*;[3] its democracy, the façade of dictatorship
(instead of monarchs we had dictators). Nor did the shift in
ideology lead to a change in social structures, and still less to
a change in psychic structures. Laws changed, but not men or
the relations of ownership and domination. During the civil
and foreign wars of the nineteenth century, the Creole aristocracy was forced out of its seats of power by mestizo groups.
The army was the school for the new groups of leaders. Backed
by military force, the regime sought and obtained the protec-
tion of foreign powers, especially of the United States. In its

3. Rule by *caciques*, petty Indian chiefs. (Like *caudillismo*, the term, by extension, has come to mean political bossism.) (TRANS.)

first phase, during the second half of the nineteenth century, the imperial career of the republic of the United States coincided with its backing of the liberal regime in Mexico, which soon became a dictatorship. This phenomenon, *mutatis mutandis*, has been repeated throughout Latin America. The liberal revolution, begun at Independence, did not result in the implantation of true democracy, or in the birth of a national capitalism, but in a military dictatorship and an economic regime characterized by *latifundismo*[4] and concessions to foreign companies and consortiums, particularly those in the United States. It was a barren liberalism and produced nothing comparable to pre-Columbian creations or those of New Spain: neither pyramids nor convents, neither cosmogonic myths nor poems of Sor Juana Inés de La Cruz.

Mexico went on being what it had been, but it no longer believed in what it was. The old values fell to pieces, but not the old realities, which were soon cloaked by the new progressivist and liberal values. These masked realities marked the beginning of inauthenticity and lies, endemic ills of Latin American countries. By the beginning of the twentieth century, pseudo-modernity was in full swing: railways and *latifundismo*, a democratic constitution and a *caudillo* within the best Hispano-Arabic tradition, positivist philosophers and pre-Columbian *caciques*, Symbolist poetry and illiteracy. The adoption of the U.S. model contributed to the disintegration of traditional values; the political and economic action of U.S. imperialism shored up the archaic social and political structures. This contradiction revealed that the ambivalence of the giant was not imaginary but real: the country of Thoreau was also the country of Roosevelt-Nebuchadnezzar.

The Mexican Revolution was an attempt to recover our past and to conceive a national project that would not be the negation of what we had been. But I am doing something worse

4. The holding of large landed estates (*latifundios*) by individuals. (TRANS.)

than being inaccurate—I am oversimplifying when I speak of the Mexican Revolution as though it were *one*. From the very beginning it was broken up into a number of contradictory movements, which, rather than representing ideological programs and political philosophies, were popular reactions, spontaneous uprisings around a leader. The word "revolt" would be more appropriate than "revolution."

Among the revolutionary groups was one that, instinctively, was aimed at correcting the course adopted by our groups of leaders since Independence: the peasant movement headed by Emiliano Zapata. What the Zapatistas demanded and really wanted was a return to origins, to a type of precapitalist agrarian society: the self-sufficient village, characterized by communal ownership of the land and a social, economic, and spiritual organization in which the basic unit was not the individual but the family. The Zapatistas bore as their standard an image of the Virgin of Guadalupe, the same standard around which the barefoot peasants who had fought for Independence had rallied. The image of the Virgin admirably symbolized not the march toward progress and "modernity," but the return to roots. "Giving the land back to the villages": this phrase, the heart of Zapata's program, points to the real meaning of his movement; it aimed at returning to a situation—in part a historical reality and in part a millenarian myth—that was the very negation of the program of "modernization" of liberalism and its heir, the regime of Porfirio Díaz. The other revolutionary tendencies, by contrast, were determined to continue the work of "modernization" of the liberals and positivists, though with different methods.

The winning faction drew up a program that endeavored to harmonize the various aspirations of the revolutionaries. More than a synthesis, it was a compromise. Civic altars were erected to Zapata—Mexico is, par excellence, the land of official and bureaucratic art—but the program of "modernization," under various names, was made the central dogma of the regime that has governed us for more than fifty years. What happened later on is a familiar story: the Mexican Revolution was taken over by a political bureaucracy not without similarities to the communist bureaucracies of Eastern Europe, and by a capitalist class made in the image and likeness of U.S. capitalism and dependent upon it. In contemporary Mexico, beyond a few of us eccentrics who mistrust "development" and would like to see a change of orientation in our society, both the factions on the right and those on the left, though irreconcilable, are fellow members of the suicide cult of progress.

For a Mexican, to travel through the United States is to enter the giant's castle and visit its chambers of horrors and marvels. But there is one difference: the ogre's castle leaves us wonderstruck by its archaism, the United States by its novelty. Our present is always just a little behind the real present, whereas its present is a little bit ahead. Its present is one in which the future is already written; ours is still tied to the past. I am wrong to use the singular when I speak of our past: we have any number of them, all alive and all continually at war with one another within us. Aztecs, Mayas, Otomis, Castilians, Moors, Phoenicians, Galicians: a jungle of roots and branches that suffocates us. How to live with all of them without being their prisoner? This is the question we ask ourselves endlessly, without having yet arrived at any definite answer. We may not have found a way to accept our past, but neither have we found a way to criticize it. The difficulty Americans have had is precisely the opposite: their nation was born as a categorical criticism of the past. This act of criticism was a no less categorical affirmation of the values of modernity, such as they had been defined first by the Reformation and later by the Enlightenment. It is not that Americans do not have a past but that their past was oriented toward the future.

The conquest of the future is the United States' tradition, therefore a tradition of change, whereas the Hispanic tradition is one of resistance to change. Spain and its works: lasting

constructions and eternal, timeless meanings. To us, "valu-
able" is a synonym for "enduring." The pre-Columbian heritage accentuates this inclination: the pyramid is the very image
of immutability. The polar opposites that exist between Americans and Mexicans are epitomized in our attitudes toward
change. To us the secret lies not in getting ahead but in man-
aging to stay where we already are. It is the opposition between the wind and the rock. I am speaking not of ideas and
philosophies but of beliefs and unconscious mental structures;
whatever our ideology, even if it is progressivist, we instinctively relate the present to the past, whereas Americans relate
it to the future. Mexican workers who emigrate to the United
States have shown a remarkable capacity for *not* adapting to
American society, a capacity that has its roots in an insensitivity to the future. In them the past is alive. It is the same past
that has preserved the Chicanos, the minority group in the
United States that seems to have been able to keep its identity.
In Mexico it has not been the professionals of anti-imperialism
who have put up the strongest resistance, but the humble folk
who make pilgrimages to the Sanctuary of the Virgin of Guadalupe. Our country survives thanks to its traditionalism.

The tradition that founded Mexico and the other countries
of Spanish America began to reveal its insufficiencies and its
limitations as far back as the eighteenth century. Because the
Spanish Empire was not able to change, it broke up into so
many fragments. Today the tradition that the United States
represents is threatened with the same fate. The ideas that for
more than two hundred years have constituted modernity and
what may be called the *tradition of the future* have lost a large
part of their universal prestige; indeed, there are many who
question their self-consistency and their value. Progress was an
idea no less mysterious than the will of Allah to Moslems or
the Trinity to Catholics, yet it stirred souls and wills for two
centuries. Today we ask ourselves: progress toward what and
for what?

It is pointless to deal at greater length with the symptoms of what has been called, for almost fifty years now, "the crisis of Western civilization." During the last decade this crisis has manifested itself in an acute form in the richest, most prosperous, and most powerful nation in our world, the United States. It is not, essentially or basically, an economic or a military crisis—although its affects the economy and the global strategy of the United States—but a political and moral crisis. It is a doubt as to the path that the nation has taken, as to its goals and the methods of reaching them; it is a critique of the ability and the honesty of the men and the parties that administer the system; and, finally, more than a questioning, it is a passing of judgment on principles that have been the base and the justification of U.S. society.

Many of the problems confronting the United States, though far from insignificant, are more symptoms than causes of the ills that it is suffering. Such is the case, to cite a notable recent example, of the rebellion of young people in the sixties. The racial problem, on the other hand, affects the life of the country in a more profound and permanent way; it is a festering wound, a permanent focus of infection. The United States is faced with a dilemma: either the perpetuation of civil strife, or the construction of a multiracial society. It is not mere wishful thinking to predict that the United States will choose the latter alternative. In point of fact, this alternative has already been chosen, and American society is heading that way, though not without many setbacks and much sidetracking. Other oppositions, present since the birth of the nation, are basic ones. All societies have within them a life principle that at the same time is a death principle, a principle necessarily dual in nature; at moments of crisis it assumes the form of a contradiction. This principle involves questions of life or death, such as the wars and rivalries between cities were for the Greek *polis*, or finding a policy to deal with Christianity and the Gnostic sects was to the Roman emperors of the third and fourth centuries. The contradiction of the United States—what gave it life and may cause its death—can be summed up in a pair of terms: it is at once a plutocratic democracy and an imperial republic.

The first contradiction affects the two notions that were the axis of the political thought of the Founding Fathers. Plutocracy provokes and accentuates inequality; inequality in turn makes political freedoms and individual rights nothing more than illusions. Here Marx's criticism went straight to the heart of the matter. Since U.S. plutocracy, unlike the Roman, admittedly creates abundance, it is able to lessen and lighten the burden of unjust differences between individuals and classes. But it has done so by shifting the most scandalous inequalities from the national scene to the international: the underdeveloped countries. There are those who think that this international inequality could also be, if not entirely eliminated, at least reduced to a minimum. Recent history, however, argues against this hypothesis. But even if it were to turn out to be true, an essential point is being forgotten: money not only oppresses; it also corrupts. And it corrupts rich and poor alike. On this score the moralists of antiquity, especially the Stoics and the Epicureans, knew more than we do. U.S. democracy has been corrupted by money.

The second contradiction, intimately linked to the first, stems from the difference between what the United States is domestically—a democracy—and what its actions abroad make it—an empire. Freedom and oppression are the opposite and complementary faces of its national being. In the same way that plutocracy begins by giving rise to inequality and ends up manacling freedom, the arms that the imperial State brandishes against its enemies abroad are, by an imperceptible process which it took the Watergate scandal to bring to light, inevitably turned into instruments that the political bureaucracy uses against the country's independent-minded citizens. The needs of empire create a bureaucracy whose specialty is espionage and other methods of intelligence used in the international power struggle; this bureaucracy in turn threatens national democracy. The first contradiction put an end to the republican institutions of ancient Rome; the second, an end to the very life of ancient Athens as an independent city.

But I am not pronouncing a death sentence upon U.S. democracy. Besides being presumptuous, that would be ridiculous. Historical analogies are useful as rhetorical tropes; they are not historical laws, but metaphors. Any and every reflection on the crisis of the republic of the United States must end in a question mark. In the first place, there is no such thing as historical determinisms. Or, rather, if such determinisms exist, we do not know what they are, nor is it likely that we ever will, since they are far too vast and complex. In the second place, societies die not from their contradictions but from their inability to resolve them. When this happens, a sort of paralysis immobilizes the social body—the thinking and deliberating centers first, then the executive arms. The paralysis is a society's response to questions to which its tradition and the assumptions of history offer no answer but silence. This was what happened in the case of the Spanish Empire. All the misfortunes of the Hispano-American peoples are distant effects of this stupor, the end product of the stubbornness, pride, and blindness that overcame the Hapsburg monarchy in the middle of the seventeenth century. The United States is faced with an entirely different situation. In the very principles on which it was founded lies, if not the answer, at least the method for finding it. This method is none other than that employed by the Puritans to scrutinize the will of God in their own consciences: soul-searching, expiation, propitiation, and the action that reconciles us with ourselves and with others.

II

Latin America and Democracy

The Antimodern Tradition

The relationship between society and literature is not one of cause and effect. The link between the two is at once necessary, contradictory, and unpredictable. Literature expresses society; by expressing it, it changes, contradicts, or denies it. By portraying it, it invents it; by inventing it, it reveals it. Though society does not recognize itself in the portrait that literature puts before it, this fantastic portrait is nonetheless real: it is that of the stranger who walks at our side from our earliest infancy and of whom we know nothing, except that he is our shadow (or are we his?). Literature is an answer to the questions that society asks itself about itself, but this answer is almost always unexpected: it answers the darkness of an era with the enigmatic brilliance of a Góngora or a Mallarmé; it answers the rational clarity of the Enlightenment with the nocturnal visions of Romanticism.

The case of Latin America is an example of the intricate, complex relations between history and literature. In the course of this century, both in Hispanic America and in Brazil, many notable works of poetry and of fiction, some of them truly exceptional, have appeared. Have there been any comparable achievements in the sociopolitical realm?

In the waning years of the eighteenth century, the best and most active Latin Americans began a vast movement of social, political, and intellectual reform. This movement has not yet ended, and it has developed in a number of different, not always compatible directions. Though it is, admittedly, a bit vague, there is one word that defines all these diverse tendencies: "modernization." At the same time that Latin American societies were endeavoring to change their institutions, customs, and ways of being and thinking, Hispano-American literature was undergoing no less profound changes. The evolutions of society and literature have had common features, but they are not parallel and have produced different results. On touching upon this subject one day, I pondered the question: is Latin American literature really modern? My answer was that it was, but in an odd way: I noted in it the absence of the critical thought that gave rise to the modern West. On this occasion I propose to explore the other half of the subject: are our present-day Latin American societies modern? And if they are not, or are modern in only a hybrid and imperfect way, why is that? My reflection, of course, has few theoretical pretensions, nor is it to be taken as advice: it is simply an opinion.

For almost two centuries now, misapprehensions about the historical reality of Latin America have been accumulating. Even the names used to designate it are inexact: Latin America, Hispanic America, Iberoamerica, Indioamerica. Each of these names leaves out a part of reality. Nor are the economic, social, and political labels that are pinned on it any more apt. The notion of underdevelopment, for example, can be applied to economics and technology, but not to art, literature, ethics, or politics. The expression "Third World" is even vaguer, a term that is not only imprecise but actually misleading: what relation is there between Argentina and Angola, between Thailand and Costa Rica, between Tunisia and Brazil? Despite two centuries of European domination, neither India nor Algeria changed language, religion, or culture. More or less the same thing can be said of Indonesia, Vietnam, Senegal, and in short, of the majority of former European possessions in Asia and Africa. An Iranian, a Hindu, a Chinese belong to civilizations that are different from that of the West. We Latin Americans speak Spanish or Portuguese; we are or have been Christians; our customs, institutions, arts, and literatures descend directly from those of Spain and Portugal. For all these reasons we are one American border of the West; the United States and Canada are the other. But we can hardly claim that we are an overseas extension of Europe; the differences are obvious, numerous, and, above all, decisive.

The first difference is the presence of non-European elements. In many Latin American nations there are strong Indian components; in others, black ones; though Uruguay, Argentina, and to a certain extent Chile and Costa Rica are exceptions. Some of the Indians are descendants of the high pre-Columbian civilizations of Mexico, Central America, and Peru; others, fewer in number, are what is left of nomad populations. Both, the former especially, have refined the sensibilities and stirred the imagination of our peoples; thus many traces of their cultures, mingled with the Hispanic, appear in our beliefs, institutions, and customs: the family, public morals, religion, popular legends and tales, myths, the arts, cuisine. The influence of black populations has also been very strong. In general, it seems to me, Indians and blacks have contributed values that are diametrically opposed: whereas the former tend toward control of the emotions and cultivate reserve and inwardness, the latter exalt the orgiastic and the corporal.

The second difference between Latin American and European culture, no less profound, stems from a circumstance that is very often overlooked: the peculiar nature of the version of Western civilization embodied by Spain and Portugal. Unlike their rivals—the English, the Dutch, and the French—the Spanish and Portuguese were dominated for centuries by Islam. But to speak of domination is misleading: the splendor of

Hispano-Arabic civilization still amazes us, and those centuries of struggle were also centuries of intimate coexistence; until the sixteenth century, Moslems, Jews, and Christians lived together on the Iberian Peninsula. It is impossible to understand the history of Spain and Portugal, as well as the truly unique nature of their culture, without taking this circumstance into account. The fusion between the religious and the political, for example, or the notion of a *crusade*, appears in Hispanic attitudes with a more intense and more vivid coloration than is the case with the other European peoples. It is not an exaggeration to see in these traits the traces of Islam and its vision of the world and of history.

The third difference, in my opinion, has been a crucial one. Among the events that ushered in the modern world we find, along with the Reformation and the Renaissance, the European expansion into Asia, America, and Africa. This movement was initiated by the discoveries and conquests of the Portuguese and the Spanish. However, very shortly thereafter, and with the same violence, Spain and Portugal closed themselves off and, retreating within themselves, rejected the modernity that was dawning. The most complete, radical, and coherent expression of this rejection was the Counter-Reformation. The Spanish monarchy identified itself with a universal faith and with a unique interpretation of that faith. The Spanish monarch was a hybrid of Theodosius the Great and Abd-er-Rahman III, the first caliph of Córdoba. (It is a pity that the Spanish kings more often imitated the sectarian policy of the former than the tolerance of the latter.) Thus, whereas the other European States tended more and more to represent nationhood and to defend its particular values, the Spanish State confused its cause with the cause of an ideology. The general evolution of society and of the different States tended toward the affirmation of the particular interests of each nation, stripping politics of its sacred character and relativizing it. The idea of the universal mission of the Spanish people, defender of a doctrine reputed to be just and true, was a survival from medieval times and Arabic culture which, grafted onto the body of the Spanish monarchy, breathed new life into it in the beginning but eventually paralyzed it. Strangest of all is that this theologico-political conception has reappeared in our day, although it is no longer identified with a divine revelation: nowadays it wears the mask of a supposed universal science of history and society. Revealed truth has become "scientific truth," and is no longer incarnated in a church and a council but in a party and a committee.

The seventeenth century was the Spanish Golden Age: Quevedo and Góngora, Lope de Vega and Calderón, Velázquez and Zurbarán, architecture and Neo-Scholasticism. It would be useless, however, to search among the great names for a Descartes, a Hobbes, a Spinoza, or a Leibniz—or for a Galileo or a Newton. Theology closed Spain's doors to modern thought, and the great century of its literature and its arts was also the age of its intellectual decadence and its political ruin.

The contrast of light and dark is even more violent in the Americas. Since Montaigne, countless pages have been written on the horrors of the Conquest. There ought to be more pages reminding us of the New World creations of Spain and Portugal as well: they were admirable. Spain and Portugal founded complex, rich, and original societies, in the image of the cities that they built, at once solid and luxurious. The rule of these viceroyalties and captaincies general was ordered along two axes, one vertical and the other horizontal. The first, hierarchical, organized society according to the descending order of social classes and groups: noblemen, commoners, Indians, slaves. The second, the horizontal axis, through a plurality of jurisdictions and statuses, linked the different social and ethnic groups, each with its distinctive features, in an intricate network of duties and rights. Inequality and orderly existence side by side: contrary and complementary principles. If those societies were not just, neither were they barbarous.

Architecture is the mirror of societies, but a mirror that shows us enigmatic images that we must decipher. The opulence and refinement of Mexico City or Puebla in the middle of the eighteenth century stand in sharp contrast to the austere simplicity, bordering on poverty, of Boston or Philadelphia. A deceptive splendor: what was a dawn in the United States was a twilight in Hispanic America. Americans were born with the Reformation and the Enlightenment—that is, with the modern world; we were born with the Counter-Reformation and Neo-Scholasticism—that is, against the modern world. We had neither an intellectual revolution nor a democratic revolution of the bourgeoisie. The philosophical foundation of the absolute Catholic monarchy was the body of thought of Francisco Suárez and his disciples of the Society of Jesus. These theologians renovated, with genius, traditional Thomism and converted it into a philosophical fortress. The historian Richard Morse has shown, with penetrating insight, that the function of Neo-Thomism was twofold: on the one hand, at times explicitly and at others implicitly, it was the ideological cornerstone of the imposing political, juridical, and economic edifice that we call the Spanish Empire; on the other, it was the school of our intellectual class and modeled their habits and their attitudes. In this sense— not as a philosophy but as a mental attitude—its influence still lingers on among Latin American intellectuals.

In the beginning, Neo-Thomism was a system of thought aimed at defending orthodox beliefs against Lutheran and Calvinist heresies, which were the first expressions of modernity. Unlike the other philosophical tendencies of that era, it was not a method for exploring the unknown but a system for defending the known and the established. The Modern Age began with a criticism of first principles; Neo-Scholasticism set out to defend those principles and demonstrate their necessary, eternal, and inviolable nature. Although this philosophy vanished from the intellectual horizon of Latin America in the eighteenth century, the attitudes and habits that were consub-

stantial with it have persisted up to our own day. Our intellectuals have successively embraced liberalism, positivism, and now
Marxism-Leninism; nonetheless, in almost all of them, whatever their philosophy, it is not difficult to discern—buried deep
but still alive—the moral and psychological attitudes of the old
champions of Neo-Scholasticism. Thus they display a paradox-
ical modernity: the ideas are today's; the attitudes yesterday's.
Their grandfathers swore by Saint Thomas and they swear by
Marx, yet both have seen in reason a weapon in the service of
a Truth with a capital *T*, which it is the mission of intellectuals
to defend. They have a polemical and militant idea of culture
and of thought: they are crusaders. Thus there has been perpetuated in our lands an intellectual tradition that has little
respect for the opinion of others, that prefers ideas to reality
and intellectual systems to the critique of systems.

Independence, Modernity, Democracy

From the second half of the eighteenth century on, slowly and
timidly, new ideas penetrated Spain and its overseas possessions. In the Spanish language we have a word that expresses
very well this sort of movement, its original inspiration, and
its limitation: *europeizar*, "to Europeanize." The renovation of
the Hispanic world, its modernization, could not spring from
the workings of principles that were our very own, that we
had elaborated by and for ourselves and successfully im-
planted, but only from the adoption of ideas from outside, those
of the European Enlightenment. Thus "to Europeanize" has
been used as a synonym for "to modernize"; years later, another word appeared that has the same meaning: *americanizar*,
"to Americanize."

Throughout the twentieth century, in the Iberian Peninsula
as in Latin America, the enlightened minorities have tried by

different means, many of them violent, to change our countries, to make the leap into modernity. For this reason the word "revolution" was also a synonym for "modernization." Our wars of independence can and should be seen from this perspective: their aim was not only to separate from Spain but also, through a revolutionary leap, to transform the new countries into truly modern nations. This is a common trait of all separatist movements, though each of them, depending on the region, has had different characteristics.

The model that inspired the Latin American revolutionaries was twofold: the American Revolution, which won the United States its independence; and the French Revolution. In fact, it may be said that the nineteenth century began with three great revolutions: those waged by the American colonies, by the French, and by the nations of Latin America. All three won a victory on the battlefield, but the political and social results were quite different in each case. In the United States the revolution brought the birth of the very first society that was wholly modern, despite the taint it bore of black slavery and the extermination of the Indians. Although the French nation suffered substantial and radical changes, the new society that emerged from its revolution, as Tocqueville demonstrated, was in many respects a continuation of the centralist France of Richelieu and Louis XIV. In Latin America, the various peoples achieved independence and began to govern themselves; the revolutionaries, however, did not succeed in establishing, except on paper, regimes and institutions that were truly free and democratic. The American Revolution founded a nation; the French Revolution changed and renewed a society; the Latin American revolutions failed to achieve one of their fundamental objectives: political, social, and economic modernization.

The French and American revolutions were the consequence of the historical evolution of the two nations; the Latin American movements were limited to the adoption of the doctrines and programs of others. I underscore the word: "adoption," not "adaptation." In Latin America the intellectual tradition that, since the Reformation and the Enlightenment, had shaped the minds and consciences of the French and American elite, did not exist; nor did there exist the social classes that corresponded, historically, to the new liberal and democratic ideology. A middle class barely existed, and our bourgeoisie had scarcely gone beyond the mercantilist stage. There had been an organic relationship between the revolutionary groups in France and their ideas, and the same thing can be said of the American Revolution; in our case, ideas did not correspond to social classes. Ideas served the function of masks; they were thus converted into an ideology, in the negative sense of that word—that is, into veils that interfere with and distort the perception of reality. Ideology converts ideas into masks: they hide the person who wears them, and at the same time they keep him from seeing reality. They deceive both others and ourselves.

Latin American independence coincides with the moment of the Spanish Empire's utter prostration. In Spain, national unity had been attained not by the fusion of the various peoples of the peninsula or by their voluntary association but, rather, through a dynastic policy of alliances and forced annexations. The crisis of the Spanish State, precipitated by the Napoleonic invasion, was the beginning of disintegration. Thus the emancipation movement of the Hispano-American nations (Brazil is a different case) must also be seen as a process of disintegration. Like a new staging of the old Hispano-Arabic story with its rebellious sheikhs, many of the revolutionary leaders took over the lands they had liberated as though these were territories they had conquered. The boundaries of some of the new nations coincided with the limits that the armies of liberation had reached. The result was the atomization of entire regions, such as Central America and the Antilles. The *caudillos* invented countries that were not viable either politically or economically and that lacked, moreover, any real national identity.

Counter to all the dictates of common sense, these have sur-
vived, thanks to the contingencies of history and the complicity between local oligarchs, dictatorships, and imperialism.

Dispersion was one side of the coin; the other was instability, civil wars, and dictatorships. On the collapse of the Spanish Empire and its administration, power fell into the hands of
two groups: economic power fell to the native oligarchs, political power to the military. The oligarchies did not have sufficient power to govern in their own name. Under the Spanish
regime, civil society, far from prospering and developing as it
had elsewhere in the West, had lived in the shadow of the
State. The focal reality in our countries, as in Spain, was the
patrimonialist system. Under this system, the head of government—prince or viceroy, *caudillo* or president—directs the State
and the nation as an extension of his own patrimony—that is,
as though it were his own household. The oligarchies, made
up of owners of large estates and traders, had lived in subor-
dination to authority and lacked both political experience and
influence on the populace. On the other hand, the ascendancy
of the clergy was enormous, as was, though to a lesser degree,
that of lawyers, doctors, and other members of the liberal
professions. These groups—the seed of the modern intellectual
class—embraced, immediately and fervently, the ideologies of
the era, some liberal and others conservative. The other force,
the decisive one, was the military. In countries without democratic experience, with rich oligarchies and poor governments, the struggle between political factions inevitably led to
violence. The liberals were no less violent than the conservatives—or, rather, they were as fanatical as their adversaries.
The endemic civil war produced militarism, and militarism
produced dictatorship.

For more than a century, Latin America has lived amid disorder and tyranny, anarchical violence and despotism. Attempts have been made to attribute the persistence of these
evils to the absence of the social classes and the economic

structures that made democracy possible in Europe and in the United States. That is quite true: we have lacked really modern bourgeoisies; the middle class has been weak and numerically small; the proletariat is recent. But democracy is not simply the result of the social and economic conditions inherent in capitalism and the industrial revolution. Castoriadis has shown that democracy is a genuine political *creation*—that is to say, a totality of ideas, institutions, and practices that constitute a collective *invention*. Democracy has been invented twice, once in Greece and again in the West. In both cases it was born of the conjunction of the theories and ideas of several generations and the actions of different groups and classes, such as the bourgeoisie, the proletariat, and other sectors of society. Democracy is not a superstructure, but a popular creation. Moreover, it is the condition, the basis, of modern civilization. Hence, among the social and economic causes that are cited to explain the failures of the Latin American democracies, it is necessary to add the one I mentioned earlier: the lack of a critical and modern intellectual current. And, finally, inertia and passivity, that immense weight of opinions, habits, beliefs, routines, convictions, received ideas and customs that forms the tradition of peoples, must not be forgotten. A century ago Benito Pérez Galdós, who had long pondered all of this, put these words in the mouth of one of his characters, a clearsighted liberal: "We see the instant triumph of the true idea over the false in the sphere of thought, and we believe that it is possible for the idea to triumph with equal swiftness over custom. Custom has been made by time, as slowly and patiently as it has made mountains, and only time, working day after day, can destroy it. Mountains are not toppled by bayonet thrusts."[1]

This rapid summary would not be complete unless mention were made of a foreign element that both precipitated disintegration and fortified tyrannies: U.S. imperialism. The frag-

---

1. *La segunda casaca* [*The Turncoat*] (1883).

mentation of our countries, the civil wars, their militarism and dictatorships were, naturally, not invented by the United States. Yet that nation bears a primordial responsibility, since it seized upon this state of affairs in order to turn a profit, to further its own interests, and to dominate. It has fostered divisions between countries, parties, and leaders; it has threatened to use force, and has not hesitated to use force every time it has seen its interests endangered; when this was to its advantage, it has backed rebellions or strengthened tyrannies. Its imperialism has not been ideological, and its interventions have involved economic considerations or reasons of political supremacy. Because of all the foregoing, the United States has been one of the greatest stumbling blocks that we have encountered in our determined effort to modernize our countries. It is tragic, because U.S. democracy inspired the fathers of our Independence and our great liberals, such as Sarmiento and Juárez. Since the eighteenth century, modernization to us has meant democracy and free institutions; the archetype of this political and social modernity was the democracy of the United States. A historical Nemesis: the United States has been, in Latin America, the protector of tyrants and the ally of the enemies of democracy.

Historical Legitimacy and Totalitarian Atheology

On attaining their independence, the Latin American nations chose the democratic-republican system of government. The Mexican imperial experiment was short-lived; in Brazil, too, a republic eventually replaced the empire. The adoption of democratic constitutions in all Latin American countries, and the frequency with which tyrannical regimes rule in those same countries, make it evident that one of the characteristic features of our societies is the divorce between legal and political reality. Democracy is historical legitimacy; dictatorship is the exceptional regime. The conflict between ideal legitimacy and the dictatorships that exist in fact is yet another expression (and one of the most painful) of the rebellion of historical reality against the schemes and geometries imposed upon it by political philosophy.

The constitutions of Latin America are excellent, but they were not specifically framed for our countries. I once called them "straitjackets"; I must add here that these "straitjackets" have been rent to pieces again and again by popular uprisings. Disorders and explosions have been the form of revenge taken by Latin American realities, or, in Pérez Galdós's terms, custom, as unyielding and weighty as a mountain and as explosive as a volcano. The brutal remedy for outbreaks of violence has been dictatorships—a deadly remedy, since it inevitably brings on new explosions. The total sterility of intellectual schemes in the face of facts corroborates that our reformers lacked both the imagination and the realism of the missionaries of the sixteenth century. Impressed by the religious fervor of the Indians, the *padrecitos* sought and found, in pre-Columbian mythologies, points of intersection with Christianity. Thanks to these bridges, it was possible to cross over from the old religions to the new. On becoming Indianized, Christianity took root and was fecund. Our reformers should have endeavored to do something similar.

The attempts to reconcile formal legitimacy with traditional reality have been few. Moreover, almost all of them have failed. The most coherent and lucid movement so inspired, the Peruvian APRA,[2] engaged in a long struggle which, though it was an exemplary contribution to the defense of democracy, in the end exhausted its revolutionary energies. Others have been

2. Alianza Popular Revolucionaria Americana, founded in 1924 by Victor Raúl Haya de la Torre to champion social and economic reforms, especially for the Indians. (TRANS.)

caricatures: Peronism, for instance, which bordered at one extreme on Italian-style fascism and at the other on populist demagogy.

The Mexican experiment, despite its faults, has been the most successful, original, and profound. It began not as a program or a theory but as an instinctive response to the absence of programs and theories. Like all true political creations, it was a collective work aimed at resolving the particular problems of a society in ruins and bled white. It was born of the Mexican Revolution, a movement that destroyed the institutions that had been created by the liberals in the nineteenth century and been turned into the mask of the dictatorship of Porfirio Díaz. The heir to the liberalism of Juárez, he had instituted a regime that was a sort of mestizo version—a combination of *caudillismo*, liberalism, and positivism—of the enlightened despotism of the eighteenth century. As happens with all dictatorships, Porfirism was unable to resolve the problem of succession, which is the problem of legitimacy: as the *caudillo* grew old, the ossified regime attempted to perpetuate itself. The response was violent upheaval. Political rebellion was almost immediately transformed into social revolt.

Once victory had been won, the revolutionaries overcame (though not without stumbling and hesitating) the temptation that besets all successful revolutions and puts an end to them: that of resolving the quarrels between factions by giving a revolutionary Caesar dictatorial powers. The Mexicans managed to avoid this danger, without falling into anarchy or civil war, thanks to a twofold compromise: the limitation of the presidency to one term in office closed the door to *caudillos*; and the formation of one party grouping together labor unions, peasant organizations, and the middle class assured the continuity of the regime. The party was not, and is not, an ideological party, nor does it hold to an orthodoxy; neither is it a "vanguard" of the people or a chosen corps of militants. It is an open, rather amorphous organization led by a political bureaucracy drawn from the popular and middle classes. Thus, for more than half a century, Mexico has been able to escape that vicious circle that consists of going from anarchy to dictatorship and back to anarchy. The result has not been democracy, or despotism, but a peculiar regime, at once paternalistic and popular, which little by little—and not without setbacks, outbreaks of violence, and relapses—has steadily advanced in the direction of freer and more democratic forms. The process has been too slow, and it has been evident for a number of years that the system is becoming outworn. After the crisis of 1968, the regime undertook, realistically and prudently, certain changes that culminated in the current political reform. Unfortunately, the independent and opposition parties, apart from being clearly a minority, lack cadres and programs capable of replacing the party that has been in power for so many years. The problem of succession is again presenting itself as it did in 1910: if we do not want to expose ourselves to very serious eventualities, the Mexican system must renew itself through an internal democratic transformation. Nevertheless, this is not enough. The central and most urgent question in Mexico is to achieve a political reform that assures, once and for all, the rotation in power of the different parties through free elections. I cannot discuss the subject at greater length here. I have devoted a number of essays to it, collected in *El ogro filantrópico*, and I refer my readers to them.

The history of Latin American democracy has not been one of failure alone. For a long period the democracies of Uruguay, Chile, and Argentina were exemplary. All three of them, one after another, have fallen, and military governments have taken over. Colombian democracy, unable to solve social problems, has become immobilized and is now a mere formalism; Peruvian democracy, by contrast, has been renewed and strengthened since the military regime ended. But the most encouraging examples are those of Venezuela and Costa Rica: two true democracies. The case of little Costa Rica, in the heart of strife-torn, authoritarian Central America, has been and is admirable.[3]

To conclude this rapid summary: it is significant that the frequency of military coups d'état has never obscured the principle of democratic legitimacy in the awareness of our peoples. Its moral authority has never been challenged. Hence, invariably, on taking over power, all dictators solemnly declare that their rule is provisional and that they are prepared to restore democratic institutions the moment circumstances permit. They very seldom keep their promise, it is true; but this does not matter. What strikes me as revealing and worth stressing is that they feel obliged to make the promise. This is a phenomenon of major importance, the meaning of which very few have pondered: until the second half of the twentieth century, no one dared challenge the proposition that democracy represents historical and constitutional legitimacy in Latin America. Our nations were democratic by birth, and, despite crimes and tyrannies, democracy was a sort of historic act of baptism for our peoples. The situation has changed in the last twenty-five years, and this change calls for comment.

Fidel Castro's movement stirred the imagination of many Latin Americans, particularly students and intellectuals. He appeared as the heir to the great traditions of our peoples: the independence and unity of Latin America, anti-imperialism, a program of radical and necessary social reforms, the restoration of democracy. One by one these illusions have vanished. The story of the degeneration of the Cuban Revolution has been recounted a number of times, among others by such direct participants in the revolution as Carlos Franqui, so I shall not repeat the details yet again. I shall merely note that the unfortunate involution of the Castro regime has been the result of a con-

3. As I anticipated earlier in the section "The Latin American Perspective" above, we have witnessed a return to democracy in Argentina, Brazil, and Uruguay, as well as a revitalization of democratic institutions in Colombia. It is safe to predict the same in Chile.

catenation of circumstances: the very personality of the revolutionary leader, who is a typical Latin American *caudillo* in
the Hispano-Arabic tradition; the totalitarian structure of the
Cuban Communist Party, which was the political instrument
for the imposition of the Soviet model of bureaucratic domination; the insensitivity and obtuse arrogance of Washington,
especially during the first phase of the Cuban Revolution, before it was taken over by the communist bureaucracy; and finally, as in the other countries of Latin America, the weakness
of our democratic traditions. This last circumstance explains
why, even though its despotic nature becomes more palpable
and the failures of its economic and social policy more widely
known with each passing day, the regime still preserves part
of its initial ascendancy among young university students and
certain intellectuals. Others cling to their illusions out of des-
peration. This is not rational but it is explainable: the word
"wretchedness," in the sense of moral misery and also in the
material sense of extreme poverty, must have been invented
to describe the situation in the majority of our countries.
Moreover, among Castro's adversaries are many who are determined to perpetuate this terrible situation. Symmetrical
enmities.

It is not difficult to understand why the Castro regime still
enjoys a certain credit among certain groups. But to explain is
not to justify, much less excuse, especially when among the
"believers" we find writers, intellectuals, and high government
officials, as is the case in France and Mexico. Because of their
culture, their access to information, and their intelligence, these
persons are, if not the conscience of their peoples, their eyes
and their ears. All of them have voluntarily chosen not to see
what is happening in Cuba and not to hear the laments of the
victims of an iniquitous dictatorship. The attitude of these groups
and these individuals is no different from that of the Stalinists
of thirty years ago; some of them will one day be just as
ashamed of themselves as some of the latter have been of what
they said and what they refrained from saying.

In any event, the failure of the Castro regime is evident and undeniable. It is notable in three cardinal areas. The international: Cuba continues to be a dependent country, though it is now the Soviet Union that holds sway over it. The political: the Cubans are less free than they were before. The economic and social: its population is experiencing worse shortages and undergoing more hardships than twenty-five years ago. The accomplishments of a revolution are measured by the transformations it brings about; among them, the change in economic structures is of prime importance. Cuba was a country characterized by the monoculture of sugar, the essential cause of its dependency on the outside world and of its economic and political vulnerability. Cuba today is still dependent on sugar.

For years and years, Latin American and many European intellectuals refused to listen to the Cuban exiles, dissidents, and victims of persecution. But it is impossible to hide the truth. Just a few years ago, the whole world was stupefied by the flight of more than a hundred thousand people, an enormous figure if we consider the total population of the island. We were even more amazed when we saw the refugees on movie and television screens: they were neither bourgeois partisans of the old regime nor political dissidents, but humble folk, men and women of the people, starved and desperate. The Cuban authorities pointed out that not all these persons had "political problems," and there was an element of truth in their assertion: that mass of humanity was not made up of opponents of the regime but of *escapees*. The mass exodus of the Cubans was not essentially different from the flights of Cambodian and Vietnamese, and was motivated by the same cause. It was one of the social and human consequences of the establishment of the bureaucratic dictatorships that have usurped the name of socialism. The victims of the "dictatorship of the proletariat" are not the bourgeois but the proletarians. Like bright sunlight suddenly breaking through heavy clouds, the flight of the hundred thousand dissipated the lies and the illusions that the reality of Cuba did not allow us to see. For how long? Our contemporaries, like the mythical Hyperboreans, have a decided preference for living amid moral and intellectual fogs.

I have already pointed out that Latin American dictatorships consider themselves to be exceptional, provisional regimes. None of our dictators, not even the most brazen of them, has ever denied the historical legitimacy of democracy. The first regime to have dared to proclaim a different sort of legitimacy was Castro's. The foundation of his power is not the will of the majority as expressed by free and secret vote, but a conception that, despite its scientific pretensions, bears a certain resemblance to the Mandate of Heaven of ancient China. This conception, fabricated out of bits and pieces of Marxism (both the true variety and the apocryphal ones), is the official credo of the Soviet Union and of the other bureaucratic dictatorships. I shall repeat the hackneyed formula: the general, ascendant movement of history is embodied in a class, the proletariat, which hands it over to a party, which delegates it to a committee, which entrusts it to a leader. Castro governs in the name of history. Like divine will, history is a superior authority, immune to the erratic and contradictory opinions of the masses.

It would be pointless to try to refute this conception: it is not a doctrine but a belief, and a belief incarnated in a party whose nature is twofold: it is both a church and an army. The apprehension we feel in the face of this new obscurantism is essentially no different from that felt by our liberal forebears in the face of the ultramontanists of 1800. The dogmatists of yesteryear saw in monarchy a divine institution and in the monarch an elect of the Almighty; today's see in the party an instrument of history, and in its leaders, history's interpreters and spokesmen. We are witnessing the return of absolutism, disguised as science, history, and dialectics.

Nonetheless, there are profound differences underlying the apparent similarity between contemporary totalitarianism and the absolutism of old. In this essay I cannot explore them or dwell on them at greater length. I shall limit myself to mentioning the fundamental one: the absolute monarch exercised power in the name of a superior and supernatural authority, God; in totalitarianism, the leader exercises power in the name of his identification with the party, the proletariat, and the laws that govern historical development. The leader is universal history in person. The transcendent God of the theologians of the sixteenth and seventeenth centuries descends to earth and becomes "the historical process"; "the historical process" in turn becomes incarnate in this or that leader: Stalin, Mao, Fidel. Totalitarianism usurps religious forms, empties them of their content, and cloaks itself with them. Modern democracy had completed the separation between religion and politics; totalitarianism unites them once more, but they are now inverted: the content of the politics of the absolute monarch was religious; today politics is the content of totalitarian pseudo-religion. The bridge that led from religion to politics in the sixteenth and seventeenth centuries was Neo-Thomist theology; the bridge that in the twentieth century leads from politics to totalitarianism is a pseudo-scientific ideology that claims to be a universal science of history and of society. The subject is fascinating, but I must leave it and return to the particular case of Latin America. . . .[4]

The antidemocratic nature of this conception is as disturbing as its pseudo-scientific pretensions. Not only are the acts and the politics of the Castro regime a negation of democracy; so, likewise, are the very principles on which it is founded. In this sense the Cuban bureaucratic dictatorship is a real historical novelty on our continent: with it began not socialism but a "revolutionary legitimacy" aimed at taking the place of the historical legitimacy of democracy. Thus the tradition on which Latin America was founded has been broken.

---

4. The reader who is interested may read with profit the penetrating and enlightening reflections of Claude Lefort in *L'Invention démocratique* (1981).

Empire and Ideology

Since the middle of the last century, United States hegemony over this continent has been continuous and undeniable. Although denounced repeatedly by Latin Americans, the Monroe Doctrine has been the expression of this reality. In this sphere, too, the Cuban Revolution represents a radical break with the past. Nemesis intervened yet again: Washington's disdainful and hostile policy threw Castro into Russia's arms. Like a gift fallen from the heaven of history—where it is not dialectics that reigns but chance—the Russians received something that Napoleon III, Queen Victoria, and the kaiser had eagerly sought and never succeeded in acquiring: a political and military base in America. From the point of view of history, the end of the Monroe Doctrine has meant a return to the beginning: as in the sixteenth century, our continent is open to the expansion of extracontinental powers. Thus the end of the U.S. presence in Cuba was not a victory of anti-imperialism. The (relative) twilight of the supremacy of the United States means, unequivocally and primordially, that Russian imperial expansion has arrived in Latin America. We have made ourselves another battleground of the Great Powers. More accurately: we have been made one. It was not steps that we took but the accidents of history that have brought us to this pass. What can we do? Whether a great deal or a very little, the first thing is to try to think clearly and independently; hence, and above all else, not to resign ourselves to being mere passive objects.

More fortunate than Napoleon III in his Mexican adventure, the Russians did not need to send troops to Cuba or engage in combat. The situation is diametrically opposite to that in Afghanistan. Castro's government has liquidated the opposition, made up in large part of former supporters, and has forcibly dominated and silenced malcontents. The Soviet Union counts Cuba among its certain allies, united to it by shared interests,

ideology, and complicity. The Russo-Cuban coalition is diplomatic, economic, military, and political. In the world's foreign
offices and international forums, Cuban diplomacy supports
points of view identical to those of the Soviet Union; moreover, it serves and defends, cleverly and diligently, the latter's
interests among the nonaligned nations. Russia and the countries of Eastern Europe subsidize the languishing Cuban economy, though apparently their aid falls far short of what is
needed. Their military aid, on the other hand, is massive and
out of all proportion to the needs of the island. In reality, Cuban troops are a military advance guard of the Soviets and
have participated in guerrilla operations in Africa and elsewhere. It is not realistic—to say the least—to close our eyes,
as have certain governments, among them the Mexican, to the
preponderantly military nature of the Russo-Cuban alliance.

The importance of Cuba as a political base is greater still, if
at this juncture there is any legitimate reason to distinguish
between what is military and what is political. Havana has
been and is a center of agitation, propaganda, coordination,
and training for the revolutionary movements of Latin America. Nonetheless, the revolts and upheavals that shake our
continent, especially in Central America, are not the result of
a Russo-Cuban conspiracy or of the machinations of international communism, as the spokesmen of the United States
government keep stubbornly repeating. These movements, as
we all know, are the consequence of the social injustices, the
poverty, and the absence of civil freedoms that prevail in many
Latin American countries. The Soviets have not invented un-
rest: they exploit it and try to co-opt it for their own ends. It
must be granted that they almost invariably succeed in doing
so. The wrongheaded policy of the United States has also played
its role in creating the present situation. But, this being said, I
then ask myself: why is it that so many revolutionary movements that originate as generous responses to unjust and even
intolerable social conditions turn into Soviet tools? Why, when

their cause triumphs, do they slavishly copy in their countries
the totalitarian model of bureaucratic domination?

The organization and discipline of communist parties almost
always impress the revolutionary apprentice. These parties are
bodies that combine two forms of association possessed of
proven internal cohesion and ability to proselytize and fight
for a cause: the army and religious orders. In both, ideology
unites individual wills and justifies the division of labor and
the strict hierarchy. Both are schools of action and of obedience. The party, moreover, is the collective personification of
ideology. The primacy of the political over the economic is one
of the characteristics that distinguish Russian imperialism from
the capitalist imperialisms of the West—the political not only
as a strategy and a tactic, but also as a dimension of ideology.

Alain Besançon calls the Soviet Union an "ideocracy," and
the term is an accurate one: in that country ideology fulfills a
function that is similar, though at a much lower intellectual
level, to that played by theology at the court of Philip II. It is
one of the premodern traits of the Russian State that are proof
of its hybrid nature, a surprising mixture of archaism and modernity. At the same time, the pre-eminence of ideology explains the fascination that the communist system still holds for
simple minds and for intellectuals from countries that liberal
and democratic ideas have penetrated only recently and super-
ficially. The popular classes of Latin America, peasants and
workers who are traditionally and stubbornly Catholic, have
been insensible to the fascination of the new totalitarian absolutism; once they have lost their faith, intellectuals and the
petite and grande bourgeoisie, on the other hand, eagerly embrace this ideological substitute, hallowed by "science." The
vast majority of revolutionary leaders of Latin America belong
to the middle and upper class—that is, to those social groups
in which ideology proliferates.

Political ideology is not incompatible with realism. The history of fanaticisms is rife with astute and courageous leaders,

clever strategists, and skillful diplomats. Stalin was a monster, not a madman suffering from delusions. On the contrary, ideology frees us from scruples and then introduces into political relations, relative by nature, an absolute in whose name everything—or very nearly everything—is permitted. In the case of communist ideology, the absolute has a name: the laws of historical development. The translation of these laws into political and moral terms is "the liberation of humanity," a task entrusted by the workings of those same laws, in this era, to the industrial proletariat. Everything that serves that end, even crimes, is moral. And who defines the end and the means? The proletariat itself? No: its vanguard, the party and its leaders. Over forty years ago, in his reply to Leon Trotsky, the philosopher John Dewey proved this argument false. In the first place, the existence of such laws of historical development is extremely dubious, and it is even more doubtful that communist leaders are best suited to interpret and execute them. In the second place, even if these laws applied as rigorously as physical laws, how could an ethics be deduced from them? The law of gravity is neither good nor bad. No theorem prohibits murder or commands the practice of charity. One critic adds that, if Marx had discovered that the laws of historical development tend not to liberate men but to enslave them, would it be moral to fight for the universal enslavement of humanity?[5] Scientism is the mask of the new absolutism.

Trotsky did not answer Dewey, but since his death the number of believers in these laws that offer moral absolution to those who act in their name has not diminished but increased. It is not difficult to recognize the origins of this morality: it is a lay version of a holy war. The new absolute succeeds in winning the support of many consciences because it satisfies the age-old, everlasting thirst for totality that all humans suffer from. The absolute and totality are the two faces of the same psychic

5. Baruch Knei-Paz, *The Social and Political Thought of Leon Trotsky* (1978).

reality. We seek totality because it is the reconciliation of our isolated, orphan, errant being with the whole, the end of the exile that begins at our birth. This is one of the roots of religion and of love, as of the dream of fraternity and equality. We need an absolute because it alone can assure us of the certainty of the truth and goodness of the totality that we have embraced. At first, revolutionaries are united by a fraternity in which the drive for power and the struggle of interests and persons are still indistinguishable from the passion for justice. It is a fraternity ruled by an absolute, but in order to realize itself as a totality it must define itself in terms of what opposes it. Thus the *other* comes into being. This *other* is not merely the political adversary who professes opinions that are different from ours: he is the enemy of the absolute, the absolute enemy. He must be exterminated. A heroic dream, a terrible dream . . . and a horrible awakening: the *other* is our double.

The Defense of Democracy

At the beginning of 1980, I wrote a series of political commentaries for various daily papers in Latin America and Spain on the decade just ended. In the last of these articles (it appeared in Mexico City on January 28, 1980) I said:

The fall of Somoza has brought up a question that no one dares answer yet: will the new regime orient itself toward a social democracy, or will it try to establish a dictatorship of the Cuban type? The second alternative would be the beginning of a series of terrible conflicts in Central America, which would be bound to spread to Mexico, Venezuela, Colombia. . . . These conflicts will not be (and are not now) uniquely national in scope, nor can they be confined within the borders of each individual country. Because of the forces and the ideologies confronting one another, the armed struggles in Central America have an international dimension. Furthermore, like epidemics, they are contagious phenomena that no sanitary cordon can isolate. The social and historical reality of Central America does not coincide with its artificial division into six countries. . . . It would be an illusion to think that these conflicts can be isolated: they are already part of the great ideological, political, and military battles of our century.

Reality confirmed my fears. Somoza's overthrow, greeted with joy by democrats and democratic socialists of Latin America, was the result of a movement in which the entire people of Nicaragua participated. As always happens, a group of leaders who had distinguished themselves in the struggle placed themselves at the head of the revolutionary regime. Some of the measures of the new government, aimed at establishing a more just social order in a country that for more than a century had been plundered both by its nationals and by foreigners, were enthusiastically applauded. The decision not to apply the death penalty to the Somozistas also awakened people's sympathy. On the other hand, it was disappointing to learn that the elections had been postponed until 1985.[6] In the course of these last few years, the regimentation of society, the attacks on the one free newspaper, the stricter and stricter control of public opinion, militarization, the widespread espionage passed off as necessary security measures, the increasingly authoritarian language and actions of the leaders have all been signs that call to mind the process followed by other revolutions that have ended in totalitarian petrification.

Despite the friendship and the economic, moral, and politi-

6. Later Managua decided to hold them a year sooner, just before the U.S. election, perhaps with the intention of presenting Washington with a *fait accompli*. The government of the commanders thus obtained a democratic sanction. But the majority opposition party refused to take part in the election, casting more shadow than light on the democratic pretensions of the regime.

cal support that our government has lent that of Managua, it is no dark secret that the eyes of the Sandinista leaders are looking not toward Mexico but toward Havana in search of orientation and friendship. Their pro-Cuban and pro-Soviet inclinations are evident. On the international scene, one of the first acts of the revolutionary government was to vote, in the conference of nonaligned nations held in Havana in 1979, for the recognition of the regime imposed on Cambodia by the Vietnamese troops who had invaded it. Since then the Soviet bloc has been able to count on one more vote in international forums. I am quite aware that it is not easy for any Nicaraguan to forget the fatal intervention of the United States in the internal affairs of the country for more than a century, nor its complicity with the Somoza dynasty. But do past grievances, which justify anti-Americanism, justify pro-Sovietism? The government of Managua could have taken advantage of the friendship of Mexico, France, and the Federal Republic of Germany, as well as the sympathy of the leaders of the Second International, in order to explore a path of independent action that would neither deliver it into the hands of Washington nor turn the country into a bridgehead of the Soviet Union. It has not done so. Must Mexicans go on offering their friendship to a regime that chooses to have others as friends?

In *Vuelta* (no. 56, July 1981) Gabriel Zaid published an article that is the best piece of reporting I have read on El Salvador, as well as being an enlightening analysis of the situation in that country.[7] His article corroborates that the logic of terror is mirror logic: the image of the assassin the terrorist sees is not his adversary's image but his own. This psychological and moral truth is also a political one: the terrorism of the military and of the ultraright is repeated, mirror-fashion, in the terrorism of the guerrillas. But neither the junta nor the guerrilla

---

7. Published in English as "Enemy Colleagues: A Reading of the Salvadoran Tragedy," *Dissent*, Winter 1982.

forces is a homogeneous bloc; both are split into various groups and tendencies. Thus Zaid suggests that the possibility of a solution other than the extermination of one of the two sides fighting each other may lie in discovering, in both camps, groups that are resolved to turn from armed struggle to dialogue. This is not impossible: the immense majority of Salvadorans, whatever their ideology, are against violence—whether from the right or from the guerrillas—and yearn for a return to peaceful and democratic ways. The elections of March 28 have corroborated Zaid's analysis: despite the violence unleashed by the guerrillas, the people turned out and lined up in the streets for hours, exposed to rifle fire and bombs, in order to vote. It was an admirable example, and the indifference of many to this peaceable heroism is yet another sign of the baseness of the times we live in.

The meaning of this election is undeniable: the vast majority of Salvadorans side with democratic legality. The vote favored Duarte's Christian Social Party, but a coalition of parties of the right and the ultraright may cheat him of victory. This situation could have been avoided if the guerrillas had agreed to take part in the democratic confrontation; according to the *New York Times* correspondent in El Salvador, they would have received between 15 and 25 percent of the votes. A tragic abstention. If the rightists assume power, they will prolong the conflict and cause irreparable damage: regardless of whether they or the guerrillas win, democracy will be the loser.[8]

Written within the situation in Central America, as though in code, is the entire history of our countries. To decipher it is to contemplate ourselves, to read the story of our misfortunes.

8. I wrote the above two days after the elections in El Salvador. Two years later elections were held again, and the winners were Duarte and his party. The new Salvadoran government initiated conversations with the guerrillas in order to find a political solution to the conflict. The case of Salvador is exemplary. If the Nicaraguan regime really wants peace, it will likewise negotiate with its opponents.

The first of them, destined to have the most fateful consequences, was independence: in liberating us, it divided us. This fragmentation caused tyrannies to multiply, and the battles between tyrants made the meddling of the United States all the easier. Thus the crisis in Central America has two faces. On the one side, fragmentation led to dispersion, dispersion to weakness, and weakness to what is today a crisis of independence: Central America is a battleground of the superpowers. On the other, the defeat of democracy means the perpetuation of injustice and of material and moral misery, whoever the winner, colonel or commissar.

Democracy and independence are complementary and inseparable realities: to lose the first is to lose the second, and vice versa. Central Americans must be helped to win the double battle for democracy and for independence. Perhaps it is not beside the point to reproduce the conclusion of the article I referred to earlier:

The international policy of Mexico has traditionally been based on the principle of nonintervention. . . . It was and is a juridical shield, a legal weapon. It has defended us, and with it we have defended others. But today this policy is inadequate. It would be incomprehensible if our government were to close its eyes to the new configuration of forces on the American continent. In the face of situations such as those that might come about in Central America, it is not enough to set forth abstract doctrines of a negative order: we have principles and interests to defend in this region. It is a question not of abandoning the principle of nonintervention but of giving it a positive content: we want democratic and peaceful regimes on our continent. We want friends, not armed agents of an imperial power.

The problems of Latin America, it is said, are those of an underdeveloped continent, yet the term "underdeveloped" is misleading: it is not a description but a judgment. That statement says something without explaining. Underdevelopment of what, why, and in relation to what model or paradigm? It is a technocratic concept that disdains the true values of a civilization, the physiognomy and soul of each society, an ethnocentric concept. This does not mean that we should ignore the problems of our countries: economic, political, and intellectual dependence on the outside, iniquitous social inequalities, extreme poverty side by side with wealth and extravagance, lack of civil freedoms, repression, militarism, unstable institutions, disorder, demagogy, mythomania, empty eloquence, falsehood and its masks, corruption, archaic moral attitudes, machismo, backward technology and scientific lag, intolerance in the realm of opinion, belief, and mores.

The problems are real; are the remedies equally real? The most radical of them, after twenty-five years of application, has produced the following results: the Cubans today are as poor as or poorer than they were before, and far less free; inequality has not disappeared: the hierarchies are different, and yet they are not less rigid but more rigid and draconian; repression is like the island's heat: continuous, intense, and inescapable; it continues to be economically dependent on sugar, and politically dependent on the Soviet Union. The Cuban Revolution has petrified: it is a millstone about the people's neck. At the other extreme, military dictatorships have perpetuated the disastrous and unjust *status quo*, abolished civil rights, practiced the cruelest repression, succeeded in resolving none of the economic problems, and in many cases exacerbated the social ones. And, gravest of all, they have been and are incapable of resolving the central political problem of our societies: that of the succession—that is, of the legitimacy—of governments. Thus, far from doing away with instability, they foster it.

Latin American democracy was a late arrival on the scene, and it has been disfigured and betrayed time and time again.

It has been weak, hesitant, rebellious, its own worst enemy, all too eager to worship the demagogue, corrupted by money, riddled with favoritism and nepotism. And yet almost everything good that has been achieved in Latin America in the last century and a half has been accomplished under democratic rule, or, as in Mexico, a rule *heading toward* democracy. A great deal still remains to be done. Our countries need changes and reforms, at once radical and in accord with the tradition and the genius of each people. In countries where attempts have been made to change the economic and social structures while at the same time dismantling democratic institutions, injustice, oppression, and inequality have become stronger forces than ever. The cause of the workers requires, above all else, freedom of association and the right to strike, yet this is the very first thing that their liberators strip them of. Without democracy, changes are counterproductive; or, rather, they are not changes at all.

To repeat again, for on this point we must be unyielding: changes are inseparable from democracy. To defend democracy is to defend the possibility of change; in turn, changes alone can strengthen democracy and enable it to be embodied in social life. This is a tremendous, twofold task. Not only for Latin Americans: for all of us. The battle is a worldwide one. What is more, the outcome is uncertain, dubious. No matter: the battle must be waged.

III

The Contaminations of Contingency

Verbal Hygiene

With a certain regularity, languages suffer from epidemics that for years infect their vocabulary, prosody, syntax, and even their logic. Sometimes the disease contaminates the entire society; at other times, isolated groups. In the last fifty years, philosophical and critical language has been attacked by three infections: phenomenology and existentialism, Marxist sects, and structuralism. The first has now almost completely disappeared, although it has left many invalids in its wake. Whereas the other two have already passed their crisis, as physicians say, they have become encysted in wild and remote regions of the periphery, such as Latin American universities. The remedies against these ailments are well known: laughter, common sense, and, in short, mental hygiene.

This is the method that Antonio Alatorre applied, with wit and intelligence, in an essay that appeared in the December 1981 issue of the *Revista de la Universidad de México*, "Crítica literaria tradicional y crítica neo-académica." What did Alatorre reproach neo-academic criticism for? The same thing that

Roland Barthes reproached his disciples and followers for: conjuring the pleasure of the text away, like a sleight-of-hand artist. The business of the critic, Alatorre maintained, "is the ebb and flow that exists between literary pleasure and the literary experience." In a recent book (*En lisant, en écrivant*, 1981), Julien Gracq says somewhat the same thing, though more breezily:

I ask the literary critic to tell me why reading this book gives me a pleasure that another book cannot give me. . . . A book that pleases me is like a woman whose charms seduce me into an affair with her: I couldn't care less about her family, her birthplace, her class, her relations, her education, her childhood, her friends. . . . What a travesty and what an imposture the *métier* of the critic is! Being an expert in *love-objects*! . . . The truth is that there is no point in concerning ourselves with literature unless it represents to us a repertory of femmes fatales and creatures of perdition.

The idea of literature that Gracq and Alatorre share casts an ambiguous light, at once sparkling and somber, on the expression "literary pleasure": it turns into passion. It is an idea that rapidly dispels the pretentious notion that we can construct a "science of literature," for the foundations of this would be the quicksands of desire.

I confess that my reproof goes beyond the mild reproaches of Gracq and Alatorre. It seems to me that the defects of modern criticism are not only of a literary nature but intellectual and moral as well. In order to judge a work, the contemporary literary critic enlists the aid of the so-called social and human sciences; from that lofty scientific standpoint he imparts his judgments, certain that he knows more about the work in question than the author himself. Sociology places all human knowledge at his disposal; psychoanalysis and linguistics make any and every professor a combination of Aristotle and Merlin.

Gracq finds it shocking that critics view the poet and the novelist as little more than a "product, a secretion of language." He is right, but it is an even more grievous offense to condemn an artist or a thinker on the grounds that he does not think or believe as we do. Literary infection is less virulent and less damaging than ideological infection. The former stems from a reversal of the traditional perspective: it sees the author not as the creator of a language but language as the creator of an author; the latter judges authors not on the basis of what they say but on the consequences of their having said it: is it favorable or unfavorable to the interests of my party?

In Mexico we have recently had an example of how widely the ideological infection has spread and how deeply it has penetrated. Gabriel Zaid in his 1981 essay set before his readers an interpretation of the guerrilla war that for several years now has drenched El Salvador in blood. The essay—as its title ("Enemy Colleagues: A Reading of the Salvadoran Tragedy") indicates—cites a wealth of documentary proofs to show the strange and bloody symmetry that rules the acts of the factions fighting for power in this little Central American country. Zaid's interpretation provoked the indignation of all those virtuous souls who look upon the conflict in Central America as yet another episode in the cyclical combat between good and evil. Zaid's discussion was significant both for the number of those who attacked him (more than twenty) and for the number of daily papers and periodicals in which the views of his critics appeared: almost every major publication in Mexico City. A deceptive multiplicity: though the voices were many, all of them said the same thing. The differences between them were not a matter of substance but of style: some set forth, not without verve, lively arguments to counter his, and others, as usual, took advantage of the occasion to vomit up their bile and vent their spleen. The polemic soon turned into a lecture on morality and thus opened up far broader perspectives than that of the events of the moment. That is why I venture to comment briefly on the whole controversy.

The Logic of Revolutions

My first observation is this: whether we approve of Zaid's
readings of events or whether they strike us as being too narrow in scope or wrong, there is no denying that the facts on
which he bases them are certain, as even his adversaries were
obliged to admit. This is a crucial admission, for they did their
best first to hide the criminal terrorist practices of various groups
of guerrillas in El Salvador and then to minimize them. When
it was no longer possible, as the expression goes, to hide the
sun with one finger—even though that finger was the elastic
one of dialectics, which changes form and size to fit the needs
of the matter under discussion—they agreed that the facts were
certain, although the most intelligent among them, Adolfo Gilly,
tried to explain them by resorting to a phantasmagorical "logic
of history." In the case of revolutions, this logic is characterized "by the irruption of the masses in their own destiny," by
intervening directly in public life to take the place of "the specialists: monarchs, ministers, members of parliaments, journalists" (Leon Trotsky, cited by Adolfo Gilly). We had best compare
this idea with the facts.

The most cursory glance at the history of modern revolutions, from the seventeenth century to the twentieth (England,
France, Mexico, Russia, China), shows that in all of them,
without exception, from the very first days of the movement,
groups possessed of greater initiative and talent for organization than the majority, and armed with a doctrine, make their
appearance. These groups very soon separate themselves from
the multitudes. In the beginning, they listen to and follow the
multitudes; later they guide them; later still they represent them;
and eventually they supplant them.

In all revolutions, once the old regime has been overthrown,
power struggles between factions break out. These struggles

are always carried on out of sight of the people, and are waged, naturally, at the people's expense. They are not popular struggles but battles in committee meetings behind the scenes. The Jacobins liquidated the Girondins, who constituted the majority of the Convention; the faction led by Robespierre and Saint-Just liquidated in turn the factions that, to use the conventional terminology, were to the right and the left of them (Danton, Hébert). The seizure of power by the Bolsheviks followed, in broad outline, the same pattern. A minority that claimed, as do all minorities, to be acting in the name of the majority, forced the majority out and took its place: cadets, Mensheviks, leftist social-revolutionaries, anarchists. Once in power, the various Bolshevik factions, behind the backs of the masses or above them, destroyed one another in their turn, until Stalin exterminated all his rivals. Something similar happened in the course of the Mexican Revolution: assassinations of Zapata, Carranza, and Villa, before the eventual triumph of Obregón. Taking into account the inevitable local differences, the same can be said of what has happened in China and elsewhere.

The famous logic of revolutions, which, according to Trotsky, is characterized by the participation of the masses in history, appears nowhere in these episodes. If revolutions have a logic, it must be granted that it unfolds in a direction and in a sense that are precisely the reverse of those described by Trotsky. In the beginning the masses—a deplorable term—are the protagonists of events, but they are soon replaced by the sects of professional revolutionaries, with their committees, their caesars, and their secretaries general. In all cases the people have been shoved aside by extremist minorities, whether Jacobins or Thermidorians. The case of El Salvador is less faithful still to the supposed "logic of revolutions." In great revolutions—such as the French, the Mexican, or the Russian—the people intervene in the first phase, and it is they who are responsible for overthrowing the old regime (monarchy, Porfirism,[1] czarism); in a second phase, the revolutionary factions fight one another for power and annihilate one another. In El Salvador the people, *before* seizing power, showed as great a repugnance toward the extremists of the right as toward the guerrilla extremists. For a number of years now, the people have been caught between two fierce armed minorities.

The intellectuals who call themselves leftists—a term that has ceased to have any precise meaning—are deaf to these arguments. The moment a fact proves their simplistic conceptual schemes wrong, they shake their heads, smile, and accuse their opponents of being "empiricists," blind to "the complexity of the social fabric and incapable of conceiving of social phenomena in terms of totalities." Logorrhea and self-importance. It is as though a weaver, out of love for the geometry of his design, stubbornly refuses to see the gaping holes in the fabric that he is weaving. Theories serve to explain the facts, not to conjure them away, or to substitute ideological entelechies for them. When the facts disprove a theory, it is necessary to abandon it or modify it. This is what these intellectuals have failed to do.

The idea that there exists a logic of revolutions presupposes the existence of a general logic of history. Does this logic exist? Does history have a meaning? The whole idea is more than questionable. Raymond Aron states: "The modern philosophy of history begins with the rejection of Hegelianism. The ideal no longer is to determine the meaning of the future once and for all. . . . The Critique of Pure Reason ended the hope of arriving at the truth of the noumenal world; in the same way the critical philosophy of history renounced the attempt to explain the ultimate meaning of human evolution. The analysis of historical knowledge bears the same relationship to the philosophy of history that Kantian criticism bears to dogmatic metaphysics" (*La Philosophie critique de l'histoire*, 1959). It is also going much too far to say that "pragmatism excludes the

---

1. The dictatorship and policies of Porfirio Díaz, which precipitated the Mexican Revolution.

masses from politics because it excludes the class struggle from history." In the first place, it is not Zaid's criticism, pragmatic or not, that excludes the masses: it is elites, revolutionary or reactionary, that exclude them, by force of arms, even as they claim to be acting in the masses' name. In the second place, history is class struggle but it is also many other, no less decisive things: techniques and their changes, ideologies, beliefs, individuals, groups—and chance.

The idea that many Latin American intellectuals have of what a revolution is, is no less dubious. In reality, there are as many visions of revolution as there are historians. For Marxists, the French Revolution was the seizure of political power by the bourgeoisie; Tocqueville, however, demonstrates convincingly that at the end of the *ancien régime* the bourgeoisie had already forced the nobility out of key points not only in the economic system but in the State as well. For Tocqueville, Furet says, "the French Revolution, far from being a brutal rupture, completes and perfects the work of the monarchy. The French Revolution can be understood only in and through historical continuity" (François Furet, *Penser la révolution française*, 1978). The paradox is that the revolution "achieves this continuity in [the realm of] facts, whereas it appears as a rupture in [the realm of] people's awareness." I shall give another example: the Mexican Revolution, in one of its central aspects—the function of the State in the process of modernization—continues the work of Porfirism. This is, of course, not meant to deny the originality of the French Revolution in the face of the *ancien régime*, or that of the Mexican Revolution in the face of the Porfirian dictatorship. These examples simply show the complexity of historical phenomena and their resistance to summary explanations of the same general type as the simplistic "logic of history." Nor does this mean that history is a meaningless and incomprehensible process, but, rather, that historical understanding must always take into account the particularity of each and every phenomenon.

Because of their very complexity and the number of factors,

circumstances, and persons that participate in each one of them, historical facts always require plural explanations. There is never a single explanation for a historical fact, not even the simplest. The principle of causality, today viewed with reservations even in physics and the natural sciences, has always been difficult to apply in the field of history. The reason is obvious: it is practically impossible to determine all the causes that have a bearing on each fact. There is, naturally, a hierarchy of causes—some are more important than others—but the hierarchies are forever shifting. At times the principal cause is the personal element, at other times economic or ideological circumstances, or, as often happens, the appearance of the factor that by definition is unpredictable: sheer happenstance. Not to mention the point of view of the historian, which is inevitably and unavoidably personal.

Are there historical laws? The question cannot be answered with any certainty; the most we can say is that, if there are, they have not yet been discovered. We might also add: even if we could discover all the factors involved, it would be almost impossible—and, what is more, pointless—to attempt to reduce them to a law. This does not mean that history must be an eternal *terra incognita*: its complexity resists the strict formalism of the natural sciences but not *comprehension*. This word means to approach, to embrace, to understand, to penetrate—not to reduce. Furet gives an example of what we must understand by "comprehension." Within the French Revolution there were various "revolutions," among them a peasant revolution, almost totally autonomous and independent of the others (that of the aristocrats, those of the bourgeois, and that of the *sans culottes*). Unlike the others, the peasant revolution was anticapitalist. This is something, Furet states, "that scarcely fits our vision of a homogeneous revolution, opening up to capitalism and the bourgeoisie a path that had been blocked off by the *ancien régime*." The same thing happened in the Mexican Revolution: Zapata's revolution was neither that of

Madero nor that of Carranza nor that of Obregón and Calles: it was not a revolution whose aim was "progress" or "development." It could even be said that, whereas the revolution of Obregón and Calles continued the work of modernization of Porfirio Díaz, Zapata's revolution was its negation.

The Ten Commandments and History

In support of their arguments, various opponents of Zaid's brought up the example of the war in Spain: no blot was cast upon the Republican cause by the abuses and crimes committed in the zone dominated by the Republic. But the moral and political problem with which we are confronted by the acts of this or that faction cannot be reduced to the simplistic terms that Zaid has quite rightly ridiculed: if my cause is a good one, the means to defend it, even if they are crimes, are also good. The problem of the relationship between ends and means is very old and complex. It would be fatuous to try to resolve it in a brief commentary such as the present one. It is not fatuous, however, to point out that it is a subject that affects not only perpetrators and victims of such acts but witnesses as well—in other words, intellectuals. Is it legitimate for a writer to remain silent about the crimes of his party? It is precisely the example cited, the war in Spain, that may help us to answer this question.

In the book on the life of Simone Weil written by her friend and disciple Simone Pétremont, the latter recounts the following:

In those days [1938] a book by Georges Bernanos came out: *Les Grands Cimetières sous la lune.* . . . Bernanos, who lived in Majorca, had been a witness to events in the zone dominated by Franco. His natural inclinations—as a Catholic writer, a

monarchist, and an admirer of the reactionary Drumont—caused
him to sympathize with the rebellion of the Spanish generals
against the Republican government. But the reign of terror instituted by the fascists in Majorca and the great number of
senseless executions had disturbed him deeply. Bernanos's book
denounced that mad, drunken death-orgy. . . . Simone Weil
felt obliged to write him a letter in which she told him that
she had undergone, from the opposite side, a similar devastating experience. In her letter Simone relates some of the events
that she had witnessed, and although not as terribly ignomi-
nious as those denounced by Bernanos in his book, they nonetheless showed that a similar atmosphere reigned in both camps.

The attitude of Bernanos the Catholic and Simone Weil the
socialist teaches us a double lesson. The first: the ends of the
cause that we defend, however lofty they may be, cannot be
separated from the means that we employ; ends do not, and
cannot, constitute our only moral criterion. The second: denouncing the atrocities that our party commits is difficult, very
difficult, but it is the first duty of the intellectual.

In the winter 1981 issue of *Dissent*, there appeared an intelligent essay on this problem of ends and means by the American writer Lionel Abel, who rightly considers it to be the central
concern of our century. Abel cites an opinion of Walter Benjamin's. According to the German critic, there are two ways of
judging violence, one from the point of view of natural law
and the other from that of positive law: "Natural law can only
judge each positive law by the critique of its ends, whereas
positive law can only judge a new law that is in the process of
being established by the critique of its means. Justice is the
criterion of the ends that are pursued; conformity with the law
is the criterion of the means employed. Natural law endeavors
to justify the means through the justice of the ends pursued;
positive law tries to guarantee the justice of the ends by the
legitimacy of the means employed."

Benjamin's argument is more subtle than it is convincing: what happens when the means are unworthy of the ends? Abel remarks that often there is a contradiction between the means that natural law finds legitimate, in view of the end pursued, and the condemnation of those means, considered by positive law to be illegitimate owing to their violence or to other circumstances. Benjamin, while not unaware of these contradictions, maintains that, in such cases, a third point of view must be sought: a superior criterion that only the philosophy of history can give us. At this point in our century, it is not difficult to criticize Benjamin's opinion. Turning the philosophy of history into an oracle means substituting the judgment of authority for the intimate moral judgment of conscience, the very foundation of ethics. It is a sleight-of-hand trick that would have shocked Socrates and Kant alike: is it legitimate to do, not what our conscience prompts but what a superior, impersonal, and remote authority tells us to do? The reason of State thus reappears, disguised no longer as Divine Providence but as the philosophy of history.

Another reason, no less decisive, leads me to reject Benjamin's idea: bowing to the dictates of the philosophy of history is even more difficult than obeying God's command. Or, rather, it is impossible: who dares maintain today that he knows the way and the designs of history? Who can judge and condemn a fellow human being in the name of a future that no one has seen and that seems more and more uncertain to us? Religions condense their principles into codes made up of a very small number of clear, absolute, categorical precepts: thou shalt not kill, thou shalt not steal, thou shalt not covet thy neighbor's wife, etc. These commandments do not depend on any particular circumstance, since they are founded upon an eternal word, outside of time. By contrast, the word pronounced by history—if it really *says* something and is not mere "sound and fury"—is temporal, and because it is time it is relative, contingent, and contradictory. It is an obscure, unintelligible word.

But even if that word could be understood and interpreted, how to base an ethics on it? Sartre tried to found an ethics of contingency and failed. It could not have been otherwise: ethics and contingency are incompatible terms.

To sum up: let us suppose that history *says something*, that we understand what it says, and that it is possible to found an ethics on this relative word. Its precepts would be inapplicable, for the reason that history carries out its dictates first and then declares what they were. It is a language that is always *a posteriori*: Mark Antony did not know that he would be defeated at Actium, nor was Lenin certain that the sealed German train would bring him into Finland Station safe and sound.

The Assassin and Eternity

The English poet Charles Tomlinson, in a poem on the death of Leon Trotsky ("Assassin"), has shown with extraordinarily penetrating insight the fatal trap into which the fanatic who believes he possesses the secret of history inevitably falls. This poem is the best refutation I know of the fallacy that sees in history a substitute for conscience. The assassin, armed with a mountaineer's ice-axe, is standing behind his victim, who is looking over some papers. At that moment he thinks:

> I strike. I am the future and my blow
>   Will have it now. If lightning froze
> It would hover as here, the room
>   Riding in the crest of the moment's wave . . .
> And as if that wave would never again recede.

Time, for the assassin, is immobilized in this instant in which the future, on coming to pass, acquires a sort of dizzying eternity. More exactly: a *philosophical* eternity. The instant is made of the same substance as history, a substance that transcends the three times—past, present, and future—because its real name is *necessity*. The assassin identifies himself with historical necessity and becomes omniscient. A false eternity, a laughable omniscience that fade away to nothingness at the very instant they appear to come into being. The weight of the fallen man, his terrible cry, the blood that stains papers and rug, the hideous gaping wound in the head, all lead only to this:

> . . . the weight of a world unsteadies my feet
> And I fall into the lime and contaminations
> of contingency: into hands, looks, time.

The ideological assassin falls from the illusory time of the philosophy of history into real time, falls from necessity to contingency, plummets from the heights of certainty to the depths of doubt. He falls into history. . . . Are there no rules in this world where everything appears to be relative? Perhaps there is one. Simone Weil points to it in her letter to Bernanos. He thought that fear was the cause of the useless and stupid cruelties that he had seen in Majorca, but Weil did not share that idea. She believed that what Zaid calls the philosophical "license to kill" had been stronger than fear: "When temporal or spiritual authorities decide that the life of a certain category of men lacks value in and of itself, other men kill them with impunity and as a matter of course. . . . Thus the end pursued in the struggle is soon lost sight of. For this end can be defined only in terms of the common good and value of the being of humanity—and the life of humanity has lost its value." The evil is dehumanization. The slaying ground and the concentration camp are institutions always preceded by an intellectual operation that consists of stripping the other of his humanity, so as to be able to enslave him or exterminate him as though he were an animal. It is a circular operation: to deny the humanity of the other is to deny our own.

IV

Peace and Democracy[1]

When my friend Siegfried Unseld announced to me that I was to be given the Peace Prize awarded each year by the Association of Publishers and Booksellers during the Frankfurt Book Fair, my first reaction was one of mingled gratitude and disbelief: why should they have thought of me? Not because of the dubious merits of my writings but, perhaps, because of my stubborn love of literature. For all of the writers of my generation (I was born in the fateful year 1914), war has been a constant and terrible presence. I began writing, that most silent of processes, in the face of and against the noisy disputes and quarrels of our century. I wrote—and I write today—because I conceive of literature as a dialogue with the world, with the reader, and with myself, and dialogue is the opposite of the noise that denies us and the silence that ignores us. I have always thought that the poet is not one who speaks but one who hears.

My disbelief is, as I said, mingled with a very real and deep feeling of gratitude. Both have grown as I listened to the Pres-

1. Acceptance speech on receipt of the Peace Prize awarded by the Association of German Publishers and Booksellers, Frankfurt, October 7, 1984.

ident of the Federal Republic of Germany, Dr. Richard von
Weizsäcker, speak about me and my writings. It is very difficult for me to express all these emotions without seeming to
be fulsome or affected. You have been very generous. I am
truly moved and I can only say to you that I will try, for the
rest of my life, to be worthy of your words.

The first historical narrative, properly speaking, in our religious tradition is the account of Cain's slaying of Abel. With
this terrible event our earthly existence begins; what happened
in Eden happened before history. With the Fall the two offspring of sin and death made their appearance: work and war.
Our condemnation began then; history began. In other religious traditions there are accounts whose meaning is comparable. War in particular has always been viewed with horror,
even among peoples who consider it to be the expression of
the battle between supernatural powers or between cosmic
principles. To escape it is to escape our condition, to go beyond ourselves—or, rather, to return to what we were before
the Fall. Thus tradition offers us another image, the radiant
reverse of this black vision of man and his destiny: in the bosom
of reconciled nature, beneath a kindly sun and sympathetic
stars, men and women live in leisure, peace, and concord. The
natural harmony between all living beings—plants, animals,
men—is the visible image of spiritual harmony. The true name
of this cosmic concord is "love"; its most immediate manifestation is innocence: men and women go about naked. They
have nothing to hide; they are not enemies, nor do they fear
each other: concord is universal transparency. Peace was a dimension of the innocence of the beginning, before history. The
end of history will be the beginning of peace: the kingdom of
innocence regained.

This religious vision has inspired many philosophical and
political utopias. If men before history were equal, free, and
peaceful, when and how did evil come to be? Although it is
impossible to know, it is not impossible to presume that an act

of violence unleashed the blind movement that we call history. Men ceased to be free and equal when they submitted to a leader. If the beginning of inequality, oppression, and war was the domination of the many by the few, how can we not see in authority the origin and the cause of the iniquities of history? Not in the authority of this or that prince, this one benevolent and that one tyrannical, but in the very principle of authority and in the institution that incarnates it: the State. Only its abolition could end the servitude of men and war among nations. Revolution would be the great rebellion of history, or, in religious terms, the return of primordial times: the return to the innocence of the beginning, amid which individual freedoms are united in social concord.

The power of attraction of this idea—conjoining the purest moral principles and the most generous dream—has been tremendous. Two factors, however, keep me from sharing this optimistic hypothesis. The first: we are here confronted with an unverified—and, I fear, unverifiable—assumption. The second: the birth of the State, very probably, did not mean the beginning but the end of the perpetual war that afflicted primitive communities. In the view of Marshall Sahlin, Pierre Clastres, and other contemporary anthropologists, men lived as free and relatively equal beings. The basis of this freedom was the strength of each man's own two arms and the abundance of goods: the society of primitive peoples was a society of free and self-sufficient warriors. It was also an egalitarian society: the perishable nature of material goods prevented their accumulation. In these simple and isolated communities, social ties were extremely fragile and discord was a permanent reality: the war of all against all. As far back as the dawn of the Modern Age, Spanish Neo-Thomist theologians had maintained that, in the beginning, men were free and equal—*status naturae*—but that because they lacked political organization (the State) they lived in isolation, defenseless and exposed to violence, injustice, and dispersion. The *status naturae* was not a synonym for innocence: like us, the earliest men were *fallen nature*. Hobbes went further and saw in the state of nature the image, not of concord and freedom but of injustice and violence. The State was born to defend men from men.

If the abolition of the State would cause us to regress to a state of perpetual civil discord, how to avoid war? From the moment they appear on this earth, States fight one another. It is not surprising, then, that the aspiration after universal peace has at times been confused with the dream of a universal State without rivals. This is as impossible a dream as that of the suppression of the State, and a remedy that is perhaps even more dangerous. The peace that would result from the imposition of the same will on all nations, even if it were the will of impersonal law, would soon degenerate into uniformity and repetition, masks of sterility. Whereas doing away with the State would doom us to perpetual war among factions and individuals, establishing a single State would result in universal servitude and the death of the spirit. Fortunately, historical experience has repeatedly banished this chimera. There are no examples of a historical society without a State; there *are* examples, indeed many of them, of great empires that have sought universal domination. The fate of all great empires warns us that this dream is not only unrealizable but also, and above all, fatal. The beginning of empires is similar: conquest and plunder. Their end as well is similar: disintegration, dismemberment. Empires are doomed to fall apart, just as orthodoxies and ideologies are doomed to split apart into schisms and sects.

The function of the State is twofold and contradictory: it keeps peace and unleashes war. This ambiguity is in us as human beings. Individuals, groups, classes, nations, and governments, all of them, all of us, are doomed to divergence, dispute, dissension; we are also doomed to dialogue and negotiation. There is a difference, nonetheless, between civil society, composed of individuals and groups, and the international society of States. In the former, controversies are resolved by the mu-

tual will of the parties involved or by the authority of the law
and the government; in the latter, the only thing that really
counts is the will of governments. The very nature of international society stands in the way of the existence of an effective
super-State authority. Neither the United Nations nor the other
international agencies have strong enough means available to
them to keep the peace or punish aggressors. They are delib-
erative assemblies, useful for negotiating, but having the defect
of being easily turned into a theater for propagandists and
demagogues.

The power of making war or peace lies essentially with governments. It is, naturally, not an absolute power: even tyrannies must take opinion and popular sentiment into account to
a greater or lesser degree before engaging in a war. In open
and democratic societies, in which governments must periodically account for their acts and a legal opposition exists, it is
more difficult to pursue a policy that is bound to lead to war.
Kant said that monarchies are more inclined toward war than
republics, since in monarchies the sovereign considers the State
to be his property. It goes without saying that in and of itself
a democratic regime is no guarantee of peace, as is proved by
the case, among others, of Periclean Athens and revolutionary
France. Like other political systems, democracy is exposed to
the deadly influence of nationalisms and other violent ideologies. Nonetheless, the superiority of democracy in this respect,
as in so many others, seems to me undeniable: war and peace
are subjects on which all of us have not just the right but the
duty to render our opinion.

I have mentioned the adverse influence of nationalist ideologies, intolerant and exclusivist as they are on the question of
peace. These ideologies become all the more baleful when they
cease to be a belief of a sect or a party and become a pillar in
the doctrine of a Church or a State. Aspiring to the absolute—
forever unattainable—is a sublime passion, but believing ourselves to be the possessors of absolute truth degrades us: we

regard every person whose way of thinking is different from ours as a monster and a threat and by so doing turn our own selves into monsters and threats to our fellows. If our belief becomes the dogma of a Church or a State, those whose beliefs differ become abominable exceptions: they are outsiders, alien, *others*, the heterodox who must be either converted or exterminated. Finally, if there is a fusion of Church and State, as happened in other eras, or if a State, by self-proclamation, grants itself exclusive title to science and history, as has happened in the twentieth century, the notions of crusade, holy war, and its modern equivalents—revolutionary war, for instance—immediately make their appearance. Ideological States are bellicose by nature—doubly so, by virtue of the intolerance of their doctrines and the military discipline of their elites and cadres. The marriage, contrary to nature, of the cloister and the barracks.

Proselytism, almost always a concomitant of military conquest, has been a characteristic feature of ideological States from antiquity to our own times. Following World War II, through conjoined political and military means, the incorporation of the peoples of so-called Eastern Europe (a misnomer) within the totalitarian system became a *fait accompli*. The nations of Western Europe appeared to be condemned to the same fate. This has not happened: they have resisted. But, at the same time, they have been immobilized: their unparalleled material prosperity has been followed neither by a moral and cultural renaissance nor by political action at once imaginative and energetic, generous and effective. To speak frankly: the great democratic nations of Western Europe have ceased to be the model and the inspiration for the elites and the minorities of other peoples. The loss has been enormous—for the entire world and for the nations of Latin America in particular; nothing on the historical horizon of these last years of the century has been able to replace the fecund influence that European culture has, since the eighteenth century, exercised on the thought, the sensibility, and the imagination of our best writers, artists, and social and political reformers.

Immobility is a disturbing symptom that becomes acute once it is recognized that its cause is the nuclear balance of power. Peace is a reflection not of the accord between powers but of their mutual terror. The countries of the West and the East would seem to be doomed to immobility or to annihilation. Thus far, terror has saved us from holocaust. But if we have escaped Armageddon we have not escaped war: since 1945 not a single day has passed without fighting in Asia or in Africa, in Latin America or in the Near and Middle East. War has become a nomad. Though it is not my concern here to discuss any of these conflicts, I must make an exception and speak of the case of Central America, which is close to my heart, and heartbreaking; what is more, it is a matter of urgent necessity to put an end to Manichaean Greeks-versus-Trojans simplifications. The first such simplification is the tendency to see the problem as merely another episode in the rivalry between the two superpowers; the second is its reduction to a local skirmish that has no international ramifications. It is evident that the United States is backing armed groups fighting the Managua regime; it is evident that the Soviet Union and Cuba are sending arms and military advisers to the Sandinistas; it is also evident that the roots of the conflict lie deep within the past of Central America.

The independence of Hispanic America (Brazil is another case entirely) precipitated the fragmentation of the former Spanish Empire. This was a phenomenon whose meaning was quite different from that of the attainment of independence by the former English colonies of North America. We Hispanic Americans are still suffering the consequences of this breakdown: within our countries, chaotic democratic regimes followed by dictatorship; without, weakness. These ills were exacerbated in Central America, a number of tiny countries without any clear national identity (what distinguishes a Salvadoran from a Honduran or a Nicaraguan?), with little economic viability and exposed to foreign ambitions. Although the five countries (Panama was invented later) chose republican regimes, none of them, with the exemplary exception of Costa Rica, succeeded in instituting an authentic and enduring democracy. The peoples of Central America very soon fell prey to the endemic evil of our part of the world: the rule of military *caudillos*. The influence of the United States began to be felt in the middle of the last century and soon became a hegemony. The initial fragmentation, the oligarchies, the dictators at once buffoons and bloodthirsty despots were not the creation of the United States, but that country took advantage of the situation, backed tyrannical regimes, and played a decisive role in the corruption of Central American political life. Its responsibility in the face of history is unquestionable, and its present difficulties in this region are the direct consequence of its policy.

Within the shadow of Washington, a hereditary dictatorship was born and thrived in Nicaragua. After many years, the conjunction of a number of circumstances—general exasperation, the birth of a new, educated middle class, the influence of a rejuvenated Catholic Church, the internal dissension of the oligarchy, and, finally, the withdrawal of U.S. aid—culminated in a popular uprising. It was national in scope and overthrew the dictatorship. Shortly after this triumph, the case of Cuba was repeated: the revolution was taken over by an elite of revolutionary cadres. Almost all of them came from the native-born oligarchy, and the majority of them have either turned from Catholicism to Marxism-Leninism or come up with a curious mixture of both doctrines. From the outset, the Sandinista leaders sought inspiration in Cuba and have been the recipients of military and technical aid from the Soviet Union and its allies. The acts of the Sandinista regime are proof of its determination to set up in Nicaragua a bureaucratic-military dictatorship modeled on the one in Havana. The original meaning of the revolutionary movement has thus been perverted.

The opposition is not homogeneous. It has a great many

supporters in the interior, but no means of expressing itself (in
Nicaragua there is just one independent daily paper: *La Prensa*).
Another important segment of the opposition is entirely cut off
from the others, living as it does in remote, desolate regions:
the indigenous minority, who do not speak Spanish, who see
their culture and their ways of life being threatened, who have
suffered depradations and outrages under the Sandinista regime. Nor is the armed opposition homogeneous: some of those
who make up this group are conservatives (their numbers include former supporters of Somoza); others are democratic
dissidents who broke with the Sandinistas *after* the overthrow
of Somoza (they include such former Sandinistas as Commandant Pastora, Juan Robelo, Arturo Cruz, and many others who
left the movement and the government as the nature of the
regime became more and more clear); still others belong to the
indigenous minority. None of these groups is fighting to restore dictatorship. The U.S. government provides them with
military and technical aid, although, as is common knowledge,
this aid is coming under increasingly heavy fire in the Senate
and in many sectors of public opinion in the United States.

I must mention, finally, the diplomatic activity of the four
countries comprising the so-called Contadora group: Mexico,
Venezuela, Colombia, and Panama. This group is the only one
to have formulated a rational policy genuinely aimed at bringing peace by peaceful means. The efforts of the four countries
are directed toward creating the conditions necessary for an
end to intervention by foreign powers, an armistice between
the contending forces and factions, and the beginning of peaceful
negotiations. This is the first and most difficult step. It is also
indispensable. The other solution—military victory by one faction or another—would merely sow the explosive seeds of another, more terrifying conflict. I point out, lastly, that the
pacification of this entire region cannot be brought about until
it is possible for the people of Nicaragua to express their opinion in truly free elections in which all parties participate. Such

elections would pave the way for the establishment of a national government. But elections, while they are essential, are not everything. In our time the legitimacy of a government is based on the free, universal, and secret vote of the people; nonetheless, to be called "democratic" a regime must fulfill other requirements, such as the preservation of human rights and basic freedoms, pluralism, and, above all, respect for the individual and for minorities. This last condition is vital for such a country as Nicaragua, which has suffered long periods of tyranny and which harbors different racial, religious, cultural, and linguistic minorities.

There are many who will dismiss this program as an impossible dream, but it is no such thing: in the midst of a bloody civil war, El Salvador held elections. Despite the terrorist methods of the guerrillas, who did their best to scare people away from the voting booths, the overwhelming majority of the populace peacefully cast its vote. This is the second time that El Salvador has voted (the first was in 1982), and on both occasions the very high percentage of the people that voted has been an admirable example of the deep-seated democratic spirit of this people and of their civic courage. The elections in El Salvador have been a condemnation of the twofold violence that afflicts the Central American nations: that of the groups on the far right and that of the guerrillas on the far left. It is no longer possible to claim that this country is not ready for democracy. If political freedom is not a luxury for El Salvador but a vital concern of its people, why is it not an *equally* vital concern of the people of Nicaragua? Have the writers who sign manifestoes on behalf of the Sandinistas asked themselves this question? How can they approve the imposition in Nicaragua of a system that they would find intolerable in their own countries? Why can they judge admirable *there* something that would be hideous *here*?

This digression by way of Central America—perhaps too lengthy a one: kindly excuse me—confirms that the defense

of peace bears a close relationship to the preservation of democracy. I emphasize again that I do not see any direct cause-
and-effect relationship between democracy and peace: democracies have more than once been aggressors. But it is my belief
that democratic rule creates an open space favorable to the
discussion of matters of public concern and hence of questions
of war and peace. The great nonviolent movements of the recent past—Gandhi and Martin Luther King are the outstanding examples—were born and developed within democratic
societies. The peace demonstrations in Western Europe and in
the United States would be unthinkable, impossible in totalitarian countries. Hence it is a logical and political error, as well
as a moral failing, to dissociate peace and democracy.

All these reflections can be summed up in a few words: in
its simplest and most essential expression, democracy is dialogue, and dialogue paves the way for peace. We will be in a
position to preserve peace only if we defend democracy. From
this principle, in my opinion, three others follow. The first is
to pursue unremittingly any and every possibility for dialogue
with the adversary; this dialogue requires, at one and the same
time, firmness and flexibility, giving ground and refusing to do
so. The second is not to yield to either the temptation of nihilism or the intimidation of terror. Freedom is not merely a
precondition of peace but a consequence: the two are indis-
soluble. To separate them is to yield to terrorist blackmail and
in the end to lose both. The third principle is to recognize that
the defense of democracy in our own country is inseparable
from solidarity with those who are fighting for it in totalitarian
countries or under the tyrannies and military dictatorships of
Latin America and other continents. By fighting for democracy, dissidents are fighting for peace—fighting for all of us.

In one of the drafts of Hölderlin's hymn to peace on which
Heidegger wrote a famous commentary, the poet says that we
humans learned to name the divine and the secret powers of
the universe for the reason that, and from the moment that,

we realized we are a dialogue and can hear each other. Hölderlin sees history as dialogue. Yet time and time again this
dialogue has been broken off, drowned out by the din of violence or interrupted by the monologue of ranting leaders. Violence exacerbates differences and keeps both parties from
speaking and hearing; monologue denies the existence of the
other; dialogue allows differences to remain yet at the same
time creates an area in which the voices of otherness coexist
and interweave. Since dialogue excludes the ultimate, it is a
denial of absolutes and their despotic pretensions to totality:
we are relative, and what we say and hear is relative. But this
relativism is not a surrender: in order for there to be dialogue,
we must affirm what we are and at the same time recognize
the other in all his irreducible difference. Dialogue keeps us
from denying ourselves and from denying the humanity of our
adversary.

Marcus Aurelius spent a great part of his life on horseback,
waging war against the enemies of Rome. He accepted armed
struggle, but not hatred, and left us these words, which we
ought never to cease to ponder: "The moment dawn breaks,
one ought to say to oneself: I shall today meet a man who is
imprudent, one who is ungrateful, one who is treacherous, one
who is violent. . . . I am intimately acquainted with him; he
is one of my kind, not through blood or family, but because
both of us partake in reason and both of us are particles of
divinity. We were born to work together, as do feet and hands,
eyes and eyelids, upper teeth and lower." Dialogue is but one
of the forms, perhaps the highest, of cosmic sympathy.